Selim Hobart Peabody

Cecil's Book of Insects

Selim Hobart Peabody

Cecil's Book of Insects

ISBN/EAN: 9783743320758

Manufactured in Europe, USA, Canada, Australia, Japa

Cover: Foto ©berggeist007 / pixelio.de

Manufactured and distributed by brebook publishing software
(www.brebook.com)

Selim Hobart Peabody

Cecil's Book of Insects

ECIL'S

OOK OF NSECTS

BY

SELIM H. PEABODY, M.A.

CHICAGO:
CLARKE AND COMPANY.
1868.

TABLE OF CONTENTS.

ABOUT ANTS.

ABOUT BEES.

HIVE BEES MAKING AND LAYING WAX.

CARPENTER AND MASON BEES, AND THEIR CELLS.

ABOUT MOSQUITOES.

TRANSFORMATIONS OF THE MOSQUITO.

ABOUT BEETLES.

THE SCARABEUS BEETLE.

ABOUT BUTTERFLIES.

THE AMPHRISIUS BUTTERFLY, CATERPILLAR, AND CHRYSALIS.

AND there's never a leaf nor a blade too mean
To be some happy creature's palace.

Lowell.

O HAPPY living things! no tongue
Their beauty might declare:
A spring of love gushed from my heart,
And I blessed them unaware.

The Rime of the Ancient Mariner.

About Ants.

BRANCH — *Articulata* — Consisting of rings, or joints.

CLASS — *Insects* — Having bodies divided into two or three distinct parts.

ORDER — *Hymenoptera* — Having membranous wings.

FAMILY — *Formicaria* — Ant family.

DO you ever find yourself, some dreamy summer day, with nothing to do? The hot, dense rays of the July sun scorch the dry grass, glow in the burning sand, and almost hiss in the water of the idle stream. The birds hide in the dense thickets, the cattle pant in the shade, and the very dog wishes he could take his jacket off. The strawberry leaves crisp in the heat, and rest upon the ground; the corn curls its green blades, and turns blue; the portulaccas shut their cups; the pansies hang their heads; even the giant sunflower droops his broad leaves, and the cabbages perspire. It is too warm to work, or to read, or to play. The boy has exhausted all his own plans for fun, and in despair asks his mother, " What shall I do ?"

I'll tell you what to do. Find an ant-hill in some shady place, where the sun will not burn your back, lie down upon your face, and watch it. You have passed such a thousand times, without knowing what curious things could be seen there. The little fellows worked all the morning, and brought up out of that hole in the middle, all the grains of sand that you see piled around in a tiny, circular fortress. One by one they brought them out and laid them in their places. Now they are thoroughly warmed by the sun, and they are carrying them back again, into the rooms which they have excavated below. If there is a flat stone near, turn it over, and you will quite likely find a much busier crowd. A large chamber, with many winding passages running hither and thither, and connecting with each other, and with other passages underneath, has been made, like the public square and the thronged streets of an old fashioned city. It is not like the exact, right-angled, stiff, modern town, but the lanes turn in and out, and yet go on with persevering directness towards some particular spot which was not down in the original plan, although a point of much consequence. Scattered all along the thoroughfares of this stone-canopied town, and quite plenty in the grand square, are many long, round, white some-things, a little like grains of wheat. People have mistaken these things for the food of the ants, and so have written,

> "The little ant, for one poor *grain*,
>   Doth tug, and toil, and strive."

But the ants lay up no food. They need none; for as soon as the hard frosts of autumn chill them, they lie down to sleep till the spring

wakes them again. If they did lay up food, it
would not be grain, for the ant can no more
eat grain than a man can eat gold, and the ant
is not so big a fool as to hoard what he can not
use.

Others have thought that these little white
sacks are the eggs of the ants; but eggs do
not grow, and surely ants can not lay eggs that
are larger than themselves. Whatever they
are, the ants evidently think them very valuable. Away they go, over the clumps of earth,
and through the tiny streets, as if to see what
has happened, and estimate the damage. They
don't quite understand it, but they are agreed
that one thing is to be done forthwith — these
precious little sacks must be carried in, out of
danger. So each grasps the nearest, and drags
it away to the hole in the centre, the gateway
of the inner town, where you see the throng
coming out. The sack is larger than the ant,
but he seizes it resolutely, and raises it over
his head. Away he creeps, but it strikes
that block of sand at the street corner, and

he can not lift it over. He lays it down and pulls the end of it round; that obstacle is past, but another is beyond. A second worker comes, and the two, by pulling at one end and lifting at the other, have brought it to the gate. Surely they can not get it through that narrow and crooked passage. One has gone below, and the sack shuts him from sight. The other tugs and pulls. It will not move. Yes, it does; see that end rise in the air; now it sinks. in the hole; now it is out of sight. But here comes another, and another. All are hurrying to the numerous stairways to the city below, and in a short time all will have vanished.

These sacks contain the young ants. The eggs were laid by the queen, and hatched by the warmth of the hot grains of sand. The grubs were fed, and grew, and finally shut themselves up in the sacks, as the caterpillar spins a cocoon, or the beetle-grub sheds his coat and becomes a chrysalis. Then the ants take great care of these sacks. They are very precious to them because they contain their children. If the air is damp and cold, or the rain falls, they carry them down into the lower rooms, and keep them warm. If the sun is warm and bright, they are brought where the warmth may be felt, without making them too dry. If they happen to be exposed, we have seen how they are hurried to a place of safety. If you should carefully dig down into the earth, you would find the underground city very extensive, the long, winding galleries lying tier after tier beneath each other, and leading to large apartments, where the ants and their children find room.

Three kinds of ants come out of these cocoon-sacks. There are males, which have four wings; females, which are much larger, and have two wings; and a third kind, called workers, or nurse-ants, which have no wings. After midsummer the several kinds may often be seen very busy about an ant hill, the winged ants trying to get away, and the workers bringing them back as often as they can find them.

The males seem to be worthless fellows, and soon disappear. They have no sting to protect themselves with, and no jaws to help them get a living.

Some of the females are caught by the workers, and taken back to the nest. Others wander away with a few followers and found new colonies, while others stray away by themselves, going out into the wide world alone. When one alights, she examines the new land which she has discovered, to see if it is fit for a home. If she is satisfied, she turns back her head, bites off her wings at the shoulders, and settles down for life. Her wings carried her from her mother's house to her new home, and henceforth her journeying is ended. Then she begins to hollow out a chamber for herself. Even if she has workers with her, she continues to toil until she has laid eggs; then she is recognized and honored as a queen. If alone, she has to continue her toil until the young from her own eggs make a colony about her. The grubs, when hatched, are fed by the nurse-ants, or by the mother, with food prepared in the stomach, and the solitary insect has much to do, to find food for herself and her hungry family.

Ants eat various substances, particularly such as are juicy, or contain sugar. They kill and eat weaker insects, and they are very fond of ripe, sweet fruit. One may be sure they will always choose the best. If the pioneers can not eat the whole of some plunder which they have found, they carry away what they can, and then bring back an army to carry off the rest. They are very fond of a substance called honey-dew. Ants are often seen running up and down the trunks of trees, even when there is no fruit on the tree to tempt them. As the trees which they visit are often sickly, they are supposed to do some injury. They are not at all to blame, but are only going to their farms to look after their cattle. The leaves and tender twigs of these trees will be found to be covered with small, pale-green insects, called *Aphides*, or Plant-lice. They are often very closely packed upon the leaf or stem, and they

do harm by sucking up the juices of the growing plant. The ant comes up the tree to his
dairy farm, and strokes one of the green lice
with his feeler; the louse gives out a single
drop of clear liquid, which the ant drinks.
Then he goes to the next, and so on, milking
his cows, or gathering honey-dew. When he
has enough, he goes back to his work, digging,
building, or feeding the young ants.

The working ant does a great deal of work
in a day. M. Huber, a French naturalist, gives
an account of a single day's work of one ant.
The insect first dug in the earth a groove or
road, about a quarter of an inch deep and four
inches long. The dirt which he took out, he
kneaded into pellets, and placed on each side
of his road, to make a wall. When this road
was finished, very smooth and straight, he
found that another was wanted, and he made
that in the same manner, and about the same
size. A man, to have done as much in proportion to his size, must have dug two ditches,
each four and a half feet deep, and seventy-two

feet long; he must have made the clay into bricks, and laid them up in walls on each side of the ditches, two to three feet high and fifteen inches thick. He must have gone over it all and made it straight and smooth; and must have made it alone, in ground full of logs and stones.

The Brown Ants, *F. brunnea*, are both miners and builders. They work either at night or in damp weather, because the sunshine dries their mortar too fast. They build a house of many stories, sometimes twenty or thirty. Each story is about a fifth of an inch high, supported by many partitions and pillars. In wet weather they take the family into the upper rooms; in dry weather they occupy the middle or the lower floors. While building, they work the damp clay in their jaws until the pellets are compact, and will adhere firmly; then they press them tightly against the tops of the partitions which they have made. As fast as one row of bricks has dried, another row is added; thus they will lay a perfectly smooth and strong ceiling two inches in diameter. When these walls are finished, the rain and sun seem only to make them harder. If a stick or straw is in their way, they at once make a beam or a post of it. If a post, they cover it with mortar until it is thick and strong enough for their work. If a beam, they build their ceiling against and around it. If a room is too large, they build partitions, and divide it into smaller rooms of suitable size.

Other Ants are carpenters. They often remove so much of a log of wood as to leave it a mere honey-comb, pierced through and through in every direction with their passages. The walls between are often as thin as paper, and yet are never broken through except where one passage crosses another. They can not know how to cut so near another passage by sight, for all is done in the dark; they can not plan or measure, as a reasoning being would do; and yet they do their work with greater delicacy and accuracy than the man who reasons and measures. For some unexplained cause, the wood through which they cut is all colored black, as if the fire had passed through it.

When these black carpenters get into a dwelling, they cause a deal of trouble. They make themselves at home in the very wood-work of the house. They gnaw a way into any wooden box which they wish to explore, and will find the least crevice into the sugar-box or cake-jar. The prudent housewife puts her pot of sweetmeats in a pan of water, but if the ants know what the jar contains, they will find a way to it, even if they crawl upon the shelf above, and drop down upon it. The family may be almost exterminated, and yet, if two or three be left, with all the resources of the nest at their command, in a little time the plagues are as thick as ever. Moreover, they bite.

Some tribes of Ants are very warlike, and they make war to capture the workers of other tribes, and obtain slaves for their own commu-nities. It is said that the kidnappers are always pale or red Ants, and that the captured slaves are black. When the red Ants are about to make a foray, they send scouts to explore the ground, who afterwards return and report their success. They then march forth in regular armies. The assailed town pours out its inhabitants, and the fight begins. Head to head, foot to foot, jaw to jaw, the sable warriors defend their homes and their children, but in vain. The victory is always with the invaders. They do not drive out their conquered foes, but they break into their homes and carry away the cocoons of the workers. The red ants return in perfect order to their own city, bearing with them their living burdens. They treat the plundered young with the same care they give their own, and the ants produced from the stolen cocoons seem to work with abundant energy and good will. The inhabitants of the besieged city, knowing what result will follow the fight, often carry away many of their young. They take them to the tops of the grass stems, and hide them amid the foliage of other plants. When the raid is over, they bring them back to the nest again. Several

kinds of ants practice this kind of warfare, and
the results are too well attested by careful
observers to admit of doubt.

Although there are many kinds, and count-
less numbers of Ants in the cooler countries of
the temperate zone, they are far surpassed in
number, in size, and in venomous power, by
those found in the hot lands of the torrid zone.
Here all kinds of reptile and of insect life seem
to be extravagantly developed, and the ants are
often so numerous and so powerful as to drive
away every other living thing.

The Saüba or Coushie Ant, *Œcodoma cepha-
lotes*, lives in South America. It is often called
the Parasol Ant. Large columns may be seen
marching along, each carrying in its jaws, and
over its head, a round piece of leaf, about the
size of a dime. Many suppose that this is actually carried to keep off the heat of the sun;
but the fact is that they use the leaves to
thatch the roofs of their houses, and to keep
the loose earth from falling in. They choose
the leaves of cultivated trees, as the orange and

the coffee. When they attack a tree, they strip
it of foliage so entirely, that it often dies. Then
they march away with their plunder, and fling
it on the ground, at the nest. Another party
of workers take up the pieces, and put them
upon the roof, covering them with dirt. These
domed houses are wonderfully large, measuring
sometimes two feet in height, and forty feet in
diameter. Their underground cities are on
even a larger scale. The smoke of burning
sulphur blown into one opening has been found
to come out at another, more than two hundred
feet away.

There are three kinds of these ants: the
winged, the large headed — sometimes called
soldiers, and the workers. The large headed
are also of two sorts: one kind has a smooth
helmet, covered with horny substance, which
one can almost see through, and the other
wears a dark helmet, covered with hairs. The
business of these large-heads is not very well
understood. The smooth helmets seem to do
nothing but walk about. They do not fight;

they do not work; they do not appear to over-look those which do work. The hairy-helmets are not known to do any more. If the top of one of the mounds be taken off, a circular well will be found in the centre, into which a stick three or four feet long may be thrust, without touching bottom. Presently some of these hairy-headed fellows, each wearing one eye in the middle of its forehead, like a fabled Cyclops, will come slowly up the smooth sides of the well, to see what is wanted. But they are not very pugnacious, and may easily be caught by the fingers.

The winged ants are the perfect males and females. They come out a little after midsummer, that is in February. The females have bodies about as large as hornets, and spread their wings nearly two inches. The males are much smaller. Although hosts pour out of the nests, few remain after a day, for the birds and insect eating animals have devoured most of them. Those which escape found new colonies in spite of all the dangers which threaten to destroy them; even the art of man can not conquer them.

Among the South American Ants are several species which are classed together, and called Foraging Ants. They belong to the genus *Eciton*. They have been confounded with the Saüba Ants, just described, but their habits are quite different. The real Foraging Ant, *E. drepanephora*, is very annoying, and very useful. These insects go out from their cities in immense armies, not very broad, but often a hundred yards long. Officers march beside the column, very busy keeping their own portion of the line in order. There is an officer to about twenty privates; their white heads nodding up and down make them quite conspicuous. The pittas, or ant thrushes, always accompany these armies, picking up the Ants for their own food; but still the band goes marching on. The people know that the Ants are on the war path, and make every preparation for their reception.

In those countries, insects of every kind get

into the houses, and multiply to an extent which almost drives the inhabitants from their homes. By day they are a trouble, and by night a pest. They bite, and suck, and scratch, and sting. They crawl over the food; they hide in the bed; they fly into the lamp, and then whirl on the table; they creep into the ink; they emit horrible smells. There are centipedes which sting, and scorpions which sting. There are cockroaches of powerful size and smell, and of insatiable appetite. As for snakes and lizards, and other creeping things, they are too common to be noticed. It is of no use to fight. Your enemies are legions of numbers innumerable. But when the Foraging Ants come, the case is altered, for nothing can stand their attack. When the pittas come about, the people open every box and drawer in the house, so as to allow the ants to explore every crevice, and then they vacate the premises.

" Presently a few scouts, which form the van-guard of the grand army, approach, and seem

to inspect the house, to see if it is worthy of a
visit. The long column then pours in and dis-
perses over the dwelling. They enter every
crevice, and speedily haul out any unfortunate
creature which is hidden therein. Great cockroaches are dragged unwillingly away, being
pulled in front by four or five ants, and pushed
from behind by as many more. The rats and
mice speedily succumb to the onslaught of their
myriad foes, the snakes and the lizards fare no
better, and even the formidable weapons of the
centipedes and scorpions are overcome.

"In a wonderfully short time the Foraging
Ants have done their work, the turmoil gradually ceases, the scattered parties again form
into line, and the army moves out of the house,
carrying its spoils in triumph. When the inhabitants return, they find every intruder gone,
and to their great comfort may move about
without treading on some unfortunate creature,
or put on their shoes without knocking them
on the floor to shake out a scorpion or a centi-
pede."

But those who are accustomed to the country are careful to keep out of the way. If a man should happen to cross the column, the ants at once dash at him, climb up his legs, and bite with their powerful and poisonous jaws. His only safety is in running away until the main army is too far off to renew the attack, and then destroying those which he has brought with him. This is not easy, for the Ants have long, hooked jaws, and bite so fiercely that they may be pulled away piecemeal, leaving the jaws in the wound to be picked out separately.

Another species, *E. prædator*, marches in broad, solid mass. It is a little creature, like our common red ant, but much brighter colored, making the trunk of a tree upon which many climb look as if smeared with a blood-red liquid.

This little red ant is exceedingly venomous: its bite brings a quenchless, burning sensation, whence the Brazilians call it "fire ant." The South American Indians require their young men to undergo the ordeal of the Tocandeiros, or fire-ants, before they can be known as warriors, or recognized as braves. A pair of mittens are made of the bark of the palm tree, long enough to cover the arms above the elbows, and are filled with the Tocandeiros. The candidate for warlike honor must put his hands into these bags of living fire, and wear them while he makes the round of the village, and dances a jig at every pause. During this march he must wear a smiling face, and chant a kind of song so loud as to be heard above all the noise his companions may make upon rude horns and drums. He must not, by word, action, or look, show any sign of the torture which he endures; if he should, he will be the ridicule of his tribe, and even the maidens will refuse to know him. When the round of the village is complete, he must pause before the chief with swifter dance, and louder chant, until he falls from exhaustion, and the burning gauntlets are removed. Then he has won his right to carry a spear with his tribe.

A species, *E. legionis*, attacks the nests of

some of the large burrowing ants. They
arrange themselves for this purpose into two
bands; one set dig into the ground and take
out pellets of earth, while the others receive
the pellets and carry them away. They will
thus sink a hole ten or twelve inches, and
always succeed in opening the nest. The materials they pull to pieces and carry home, as
well as the inmates. The community is in
wonderful discipline. Each ant knows his
place, and attends to his business.

The species *E. erratica,* is blind. The eyes
of the other varieties are very small, but in the
Blind Ant they are absolutely wanting, not
showing even a trace. They have, however,
some means of knowing light from darkness,
for they are very uneasy when brought into the
light.

They are wonderful builders, constructing
long galleries through which they travel. If a
gallery be broken into, the soldiers are seen
slowly coming out, and opening their large
jaws as if they would bite something. If not

disturbed, they retire into the gallery, and soon the workers come and repair the breach. These galleries are built upon the surface of the earth, and do not penetrate the soil.

Some Ants make their nests in trees, hanging them from the boughs, like the wasps. One of these carries its abdomen in the air, hanging over its back, and has acquired the uncouth name *Crematogaster*, or "hanging-belly." Another is called by travelers the Green Ant, *Œcophylla virescens*. The name signifies a house and a leaf, and is given because it makes its hanging nest of dried leaves. When disturbed, the Ants come pattering down upon the man below like rain-drops, seeking for spots which they can wound, and having a special faculty for finding their way down the neck.

A tribe of Ants somewhat similar to the *Ecitons* of South America, is found in Africa, and is called Bashi Kouay, or Driver Ant, *Anomma arcens*. It is the dread of all animals, from the leopard to the smallest insect. It marches through the forest in lines about two inches broad, and of incredible length. One writer asserts that he has seen a column of these insects continue passing a single point, at good speed, for twelve hours. Officers march along the line and maintain order. If the advance guard come to an open place, not shaded by trees, they build a covered way, or tunnel, of dirt moistened with their saliva. If there are sticks and leaves on the ground, they fill up only the spaces which are exposed, for the direct rays of the sun kill them very quickly. If a stream crosses their path, they make a bridge of themselves, over which the whole pass. First a single Ant swings himself from the branch of a tree which overhangs the water. Then another crawls over him, and hangs from his feet. Others follow until a living chain is formed which reaches to the water, and rests upon it. Then the wind or the current wafts the free end of the chain about until it touches the opposite shore, and the crossing is complete. If one chain bridge

is insufficient, others are made alongside. It
is asserted that the bridge is even made tubu-
lar, and that the army marches through it.

When the Ants get hungry, the long line
stops marching by the flank, as soldiers would
say, that is, following each other in line, and
moves like an army in line of battle, devouring
every thing in its way. The black men run
for their lives. In a very short time a mouse,
a dog, a leopard, or even a deer, is overrun,
killed, eaten, and only the bones are left.
When they enter a house, they clear it of every
living thing. If a fowl is the victim, they dig
out the feathers by the roots, and then pull the
flesh to pieces, fastening their strong pincers
into it, and never failing to bring away the
piece.

A white hunter killed an antelope, and
brought it to a native village. In the night
he felt himself terribly bitten, and roused his
attendants. The whole village was attacked by
a column of the Bashi Kouay, which was
attracted by the smell of the meat. The

natives protected themselves by making circles of fire and standing inside. Before morning the insects had eaten every thing they could get, and had traveled on.

During the abundant tropical rains the Drivers run together and form themselves into balls, varying in size, but usually about as large as those used in the game of ball. These balls of ants float upon the water until the land appears again, and the insects can go about their business. The natives try to destroy them by making fires over and about their nests. This does not accomplish much, for the cunning ants escape before the heat becomes too great, and will be found hanging in festoons upon the neighboring trees, and crossing from one to another by their chain bridges.

These ants are black, with a tinge of red. They have enormous heads, equaling about one third of their entire length. The jaws are sharply curved, and cross each other when closed, so that if the ant has fixed itself, its hold can not be loosened unless the jaws are opened. It has no appearance of external eyes.

Dr. Lincecum has observed an Ant in Texas, which has been called the Agricultural Ant, *Atta malefaciens*. When this species has fixed its home in good dry ground, it bores a central hole, about which it raises the surface perhaps six inches, making a low mound, which gently slopes to the outer edge. If the spot be wet, the mound is raised higher, and is even fifteen òr twenty inches high. The space about the mound is carefully cleaned and smoothed like a pavement. Nothing is allowed to grow in this circle, two or three feet from the centre, except a single species of grass. This grass the ants tend with the greatest care, cutting away the weeds within and about it. It thrives under their culture, and bears a crop of seed which resembles, under the microscope, miniature rice. When ripe, it is carefully harvested, and carried into the cells, where it is cleaned of the chaff, and packed away. If the grain gets moist in damp weather, it is taken out and

dried on the first fair day, and the sound ker-
nels are carried back again ; those which have
sprouted are thrown away. Since men have
made farms in that country, and the cattle have
eaten down the ant-rice, thus spoiling their
crop, the ants have either abandoned the pas-
tures, or those communities have perished.
They may be found in places where the cattle
can not get at their crop of grain.

Dr. Lincecum is confident, after twelve years'
observation, that these ants plant the grain,
take care of it, harvest it, and keep seed for
another sowing. Each year the crop of ant-
rice is found growing about their cities, and
not a blade of any other green thing can be
found within twelve inches of this grain.

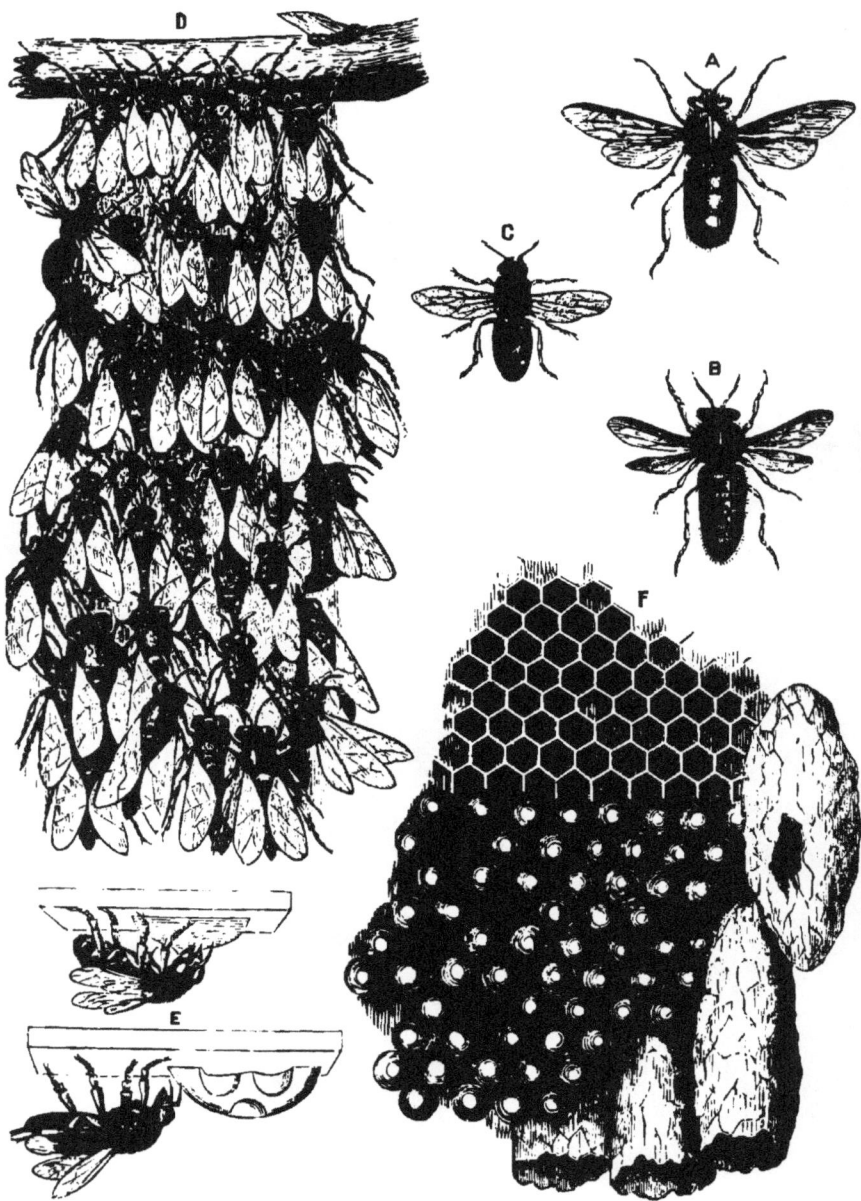

HIVE BEES MAKING AND LAYING WAX.

A, Queen Bee; B, Male; C, Worker; D, Bees clustering to make Wax; E, Bees Laying and Sculpturing Wax; F, Comb, with Empty, Full, and Queen Cells.

CARPENTER AND MASON BEES, AND THEIR CELLS.

B, Mason Bee; D, Cells of the Mason Bee; C, Carpenter Bee; A, F, Cells of Carpenter Bee; E, Comb of Humble Bees.

About Bees.

Articulata.— Insecta.

Order — *Hymenoptera* — Membrane-winged.

Family — *Apidæ* — Bee family.

IN the summer days, among the white clover heads, we find a bright, busy, buzzing Bee. He runs quickly over the round white bouquet, and thrusts his long tongue deep into every floweret. He tastes of each, and then, with cheery hum, visits another and another flower. In a little time he has gathered his sweet freight. He rises in the air, circles about for an instant, and then dashes away in the straightest of bee-lines to his home.

Another is searching the larkspur. A third is working at the snapdragon, the "frogs-mouth" of the children. Now he kicks against the lower lip of the gay corolla. It opens, and in he goes, while the door shuts after him. Presently it opens again, the Bee creeps out, goes to another, and vanishes in that. A fourth is making the round of the cucumber vines. Down he goes into the golden cup, round the sculptured pillar at the bottom, and out again, dusty with yellow pollen. He descends into a second cup, and as he rubs his way round that column, carved with a different device, he leaves a little of the golden dust to give vitality to the tiny cucumber at the base of the flower. If there were no Bees, the cucumbers and squashes would not grow.

Hear the gentle hum among the pale, graceful clusters of locust blossoms, which burden the air with their oppressive sweetness. The buckwheat field resounds with the busy murmur. They visit the honeysuckles and the morning-glories, the clematis and the violets, the lilies, the pea-blossoms, the scarlet-runners, and all the multitude of flowers that provide honey in their fragrant cups.

Before that fellow which explored the cucumber blossoms went home, he rested on a twig and scraped himself all over with his feet. He cleaned off every particle of the yellow pollen which had gathered upon his velvet coat, and put his jacket in the nicest order. Little dandy, is he? Not at all. He is only neat; and besides, the dust was partly what he came for. He kneaded it together, rolled it up carefully in a ball, and tucked it away in his trousers pocket. Not just that, either, but in a hollow inside his thigh, made on purpose for that kind of load, and lined with bristly hairs to keep the little yellow packet from falling out. One may often be seen with his two thighs loaded down, while he is still gathering his supply of honey.

Behind the house, under a little shed in the thicket of locusts and cinnamon roses, is the Bee's home. When his ancestors took care of themselves, they made their comb and stored their honey in the hollow of some old tree; they ate it themselves, unless the bears climbed the tree and took a share. Now the careful farmer provides a snug, clean box for each swarm, and pays himself from their stores.

You may stand near and watch them, if you will be quiet, and have not made yourself offensive to the Bees with some strong perfume. Their sense of smell is very acute, and many perfumes make them very cross. If you find that one begins to circle round your head with a sharp, rasping buzz, quite unlike the genial hum of those which are coming and going, and particularly if you find that two or three join in the song, and fly in the same curve, you had better go without ceremony. In an instant more you may expect them to dash in your face and sting you, and that a score of angry bees will follow their example. But you may usually approach without fear, and will find a busy community — " busy as bees."

A hive of Bees contains a queen, a few hundred drones, and may have 15,000 or 20,000 workers. The workers are those we have seen gathering honey and pollen. They are about half an inch long, nearly black, and are armed with a straight sting. The drones are about five-eighths of an inch long, and are thicker and clumsier than the workers. They have no sting. The queen is more than three-fourths of an inch long, slender and graceful; she has a curved sting.

When a swarm of Bees are newly settled in a hive, their first business is to commence building. A part clean out the hive, while most go to the fields for honey and pollen. This latter they work into a substance called propolis, with which they glue the wax to the roof of the hive, and stop up all crevices which might admit cold, or insects. The wax is produced by the Bees themselves. Those which return from the fields hang themselves from the top of the hive in bunches, festoons, ropes, and other fantastic forms, and remain quiet for about twenty-four hours. During this time the

wax exudes between the rings of the bodies of
the Bees, eight little scales coming out on
each side. One leaves the festoon, goes to the
top of the hive, and drives away the others
from the spot where it would begin. It then
takes from itself one of the scales of wax, chews
it to make it pliable, and sticks it against
the roof of the hive. When it has thus used
all its wax, another takes the place, and lays
more wax. While one works in one direction,
another works in the opposite direction. Soon
a thin partition begins to hang down, which
will separate the ends of the two rows of cells
that meet in the middle of the comb. When
the two Bees working opposite to each other
leave room between them, a third begins to cut
out a hollow in one side of the partition, and
presently two others begin to hollow on the
opposite side. As fast as the wax layers extend
the partition and make room, the sculpturers
dig out the hollows on the sides. If the reader
will press a slip of paper between the tips of
two fingers of one hand and three fingers of

the other, the paper will take the shape which
the wax partition has when the sculpturers have
followed the wax layers. The hollows made
by the ends of the fingers will represent the
bottoms of the cells on either side of the partition. Now lay a number of marbles of the
same size upon a table. They will lie most
closely if one be put down first, and six more
placed around it; when these are placed, the
others will readily find their places. If the
marbles were pressed into the surface of a sheet
of wax, they would show the arrangement of
several cells against one side of the central partition; the spaces between the marbles would
show where the partitions between cells are
made. But these spaces are triangular, and if
filled up with wax, would waste wax and space,
both which are very precious to the builders.
So they cut out all that can be spared from the
little three cornered places, and make the three
partitions meet between three cells which join
each other. Thus the six sided, or hexagonal
shape of the cells is arranged.

Now there is room for more Bees to work.
Some lengthen and widen the middle partition;
some hollow out the cell bottoms; some lay
wax for the sides of the cells, building directly
out from the central wall; some smooth the
interior of the cells. The same Bees do not lay
the wax and smooth it too. When the work
on one comb is fairly begun, the proper distance is measured, and another is laid out on
either side of the first; then two more still farther away, and so on until the ceiling is covered. In a little time all the workers find
plenty to do, and they work with such diligence
that a moderate swarm will build four thousand
cells in a day.

When the cells are made, and even before
they are finished, the queen comes to lay the
eggs. She first puts her head in the cell, as if
to see that it is properly made, then she turns
about and places an egg at the farther end.
She supplies thirty or forty cells on one side of
the comb, and then passes to the opposite side,
where she lays as many more. In this way the

grubs in the same body of comb are hatched at the same time, and the bees come out together. While the queen is laying, the workers treat her with the greatest attention. They caress her; they feed her from their own mouths; if danger threatens, they cover her with their bodies, piling up two or three inches thick. If they are pushed aside, and the queen is taken out, they seem greatly alarmed for her safety, but do not sting. Their whole anxiety is for the welfare of their beloved mistress.

The egg hangs upon the upper angle of the cell for three days. Then it bursts, and a lively little worm falls from it. At once the workers begin to look after the baby-bee. They feed it with liquid food, prepared in their own stomachs from farina, or pollen, with honey, and perhaps water. At first the liquid is quite insipid, but afterwards contains more honey. The grub eats voraciously, and the Bees bring all it can eat. They watch the brood with tender care. If a comb containing it be placed in an empty hive, they will continue to take care of it with-

out regard to other duties. By thus removing
a body of comb containing one or two queen
cells, a portion of a swarm may be transferred
to a new hive, without the usual process of
swarming.

About five days after the egg is hatched the
grub stops eating. It has nearly filled the cell,
and has curled itself into a ring. Then the
Bees seal it up in its cell with a cover of wax,
and leave it, while it spins a silken shroud like
a silkworm. This takes a day and a half; in
three days more it has changed into a pupa, or
chrysalis. First it straightens itself. Then the
parts of the perfect Bee begin to form under
the clear, white skin. The head, the eyes, the
antennæ, the wings, the feet, the rings of back
and abdomen, may all be seen under the silken
garment which seems to be laid in shining folds
about its head, and gathered up about its feet.
It looks like the living mummy of a Bee. The
skin changes from white and clear, to black
and opaque; the parts become more distinct.
On the twenty-first day from the laying of the

egg, the perfect insect throws off the black mummy wrapper, eats through the silken shroud and the wax coffin-lid, and comes forth. In half an hour she is free from the cell; she dries her wings, and on the same day goes out into the world to sip honey and gather farina with her elder sisters. As soon as the young Bee has left the cell, the workers clean it out and put it in order for another egg, or for the storage of farina or honey. A large portion of the cells are used for this purpose, the food being intended for a supply at the season when flowers are not in bloom.

The care taken of the egg and grub of the worker, though very great, can not compare with that given to the young which are to become queens. The workers act as if the fate of their nation depended upon the young creature. They feed it with a richer, more pungent, and more acid jelly, and supply more of this royal food than can be eaten. After the cell is closed up, the grub spins a cocoon, but does not complete it. This omission is often fatal to itself, but necessary to the quiet of the hive, for the queen first hatched often stings to death her rivals which have not yet appeared. If the cocoon were complete, she might not be able to pierce it, or her sting might be entangled in the silk, which would destroy her own life. The queen ceases to be a chrysalis on the sixteenth day, but she is not allowed to leave the cell until a suitable time comes. If she were to come forth while the weather was such that a swarm could not fly, there would be two queens in the same hive, and that could not be permitted. A contest would ensue, and the older and stronger would kill the younger. So the workers keep the young queen prisoner, but give her plenty to eat.

Mean while the old queen becomes agitated and impatient. She has stopped laying eggs, and runs distractedly here and there over the comb. The workers share in her excitement, and gather about her. They fly wildly about the hive, but do not go away for food. Suddenly the confused noise within ceases. In a second some workers come forth, and then the whole swarm, led by the mother queen, streams out and fills the air with a dark cloud. They hover for an instant about their old home, and then settle in a compact mass, like a ball, or bunch of grapes, upon a bush, or branch of a tree. If undisturbed they will soon fly again, and on swift wings vanish to some distant place, and probably be lost. While the swarm is quiet, they may be gathered in a bag or shaken into a hive. If the box be sweet and clean, and particularly if a little honey or wax has been rubbed in it, the Bees will almost always adopt it as their new home.

When swarming they are said to be perfectly harmless. Jardine says: " They are so intent on the acquisition of a new abode, and so anxious about the safety of their mother and queen, that what on ordinary occasions would draw forth many a vengeful weapon, now passes utterly unheeded by them; and the cultivator may lift them in handfuls, like so much grain, without in the least suffering for his boldness."

The young queens are left in the hive. After the departure of the old queen, the young one is allowed to come out of her cell. She at once goes to the other royal cells, and tries to kill the queens enclosed in them. Sometimes she succeeds, but the workers often crowd round her and hold her back. Excited by this treatment she sometimes leaves the hive, taking a quantity of workers with her, and so forms a second swarm. This may be repeated from a large hive until three or four swarms have left. It would seem that the hive must become quite deserted from such drafts upon it, but this is not the case. The many Bees which are in the field when the swarm leaves return to their old home, and there is a multitude of young Bees in the comb, which shortly come forth and supply the place of those which left.

It sometimes happens that a queen dies, and that too at a time when there are no queen grubs in the cells. Perhaps the queen has been taken away in order to see what the Bees would do. For about twelve hours every thing goes on as usual; the workers do not seem to know their loss. Then the community is in great distress. All labor is suspended. They rush in crowds to the door as if to leave the hive. They gather in groups as if consulting together. Then they seek the comb where worker grubs are hatched, and open three cells into one, making a royal cell. The one grub which is left in the cell is fed with royal jelly, and treated in every way like a queen grub. The same thing is done in three or four places, to make the result secure. The change of food, and the increased size of the cell, work a change in the larva, or produce a more complete development, and in due time it comes forth a perfect queen. It is known to be a fact that the Bees can produce a new queen for themselves if they have a comb containing grubs not more than three days old.

When a second queen is placed in a hive which has already a recognized queen, the Bees gather round the new comer, and though they do no violence, in a few hours she is either starved or suffocated. If the two queens meet, a battle follows, and one is slain. Sometimes both perish. If the Bees have lost their queen, and have discovered their loss, a new queen will be at once recognized; before the proper time has passed, they treat the new queen as if the old one were yet with them.

There is another Bee in the hive, of which little has been said. This is the drone, or male Bee. He is known by his larger size, his heavy flight, and his loud humming or droning sound. He takes no part in the work of the hive, nor does he go to the field to gather honey. His life is short. About the first of August, when the supply of honey begins to fail, the Bees seem to discover that the drones are of no use in their community, and that they can not afford to support them in idleness. The drones appear to know their danger, and cluster together in a corner. By and by the storm bursts. They are driven to the bottom of the hive, and out of doors. They have their wings bitten off. They drag two or three of their enemies with them, but their strength will not save them. They are unarmed, and the workers wear sharp, poisoned stings. Those which escape the massacre fall a prey to birds or toads, or perish with cold and hunger. So bitter is the fury of the workers, that they tear open the cells which would produce drones, kill the young, and drag the lifeless bodies out of the hive.

In all the work of the Bees, they take much pains to· keep the hive uniformly warm. In cold weather the heat comes from the clusters of their bodies, and is considerably more than that of a well warmed house. In summer the hive is cooled by ventilation. A certain number of workers may always be found in hot weather, vibrating their wings on the alighting board before the door of the hive. Inside, a still larger number is employed in the same way. They stand on the floor of the hive in lines, which separate to allow the workers to pass, and extend to the spaces between the combs. The beating of their wings forces a constant current of fresh air into the hive. This is one cause of the hum which constantly resounds from a hive where the bees are at work.

The honey may be taken from the hive, after the Bees have been removed by driving, or by suffocation, or it may be procured in extra boxes. Formerly, a dense smoke was made, the hive placed over it, and the Bees destroyed. Or the hive may be turned up, and an empty one placed over it; a few smart taps on the lower hive will drive the Bees into the upper one. But the best plan is to have the hives made in two stories, and to put suitable boxes into the upper story, communicating with the lower by holes through the ceiling. The Bees fill the boxes with comb and honey, and then they may be removed and others put in the place.

Bees are kept in most countries, but the varieties differ considerably. Fifteen or twenty kinds of hive Bees are named.

In Africa, in Australia, and in America, they are often found wild. Bee hunters sometimes derive considerable profit from the honey which they find in the hollow trunks of decayed trees. The hunter catches a Bee which is about ready to go home, marks it with a little red paint, or sticks a bit of white down to it, and then watches its flight. He goes a little distance, and takes another, which he treats in the same way. By observing several, he traces their lines to the tree, cuts it down, and obtains the honey. The wild Bees of America were not originally natives. They were brought from Europe by the English, and a swarm was carried over the Alleghany mountains in 1670 by a hurricane. The Indians call them "English Flies," and they say that the Indian and the buffalo flee before the Bees. Longfellow's Indian says of the Bees and the white clover:

" Wheresoe'er they move, before them
    Swarms the stinging fly, the Ahmo,
    Swarms the Bee, the honey-maker;
    Wheresoe'er they tread, beneath them
    Springs a flower unknown among us,
    Springs the White Man's Foot in blossom."

" Wise in their government, diligent and active in their employments, devoted to their young and to their queen, the Bees read a lecture to mankind that exemplifies their oriental name, Deburah, *she that speaketh.*"

The great family of Bees may be divided into two classes: those which live in communities, and are called Social Bees, and those which living and working alone, are called Solitary Bees. The varieties of both classes are very numerous. More than two hundred and fifty species are known in Great Britain alone. The most noted of the social Bees are the common Hive or Honey Bees, which have already been described. Another kind, familiar to all my readers, is the Humble Bee. In New England these Bees are known to boys as Bumble

Bees, or Bum-bees. In different parts of Old England they are called Foggies, Dumbledores, or Hummel-bees. Let us observe the annual circuit of a family of these Bees.

In autumn the workers, the males, and all the old females, die. The young females find some sheltered place, in moss, dead leaves, or decayed wood of an old tree, where they may pass the winter. As the cold begins they become torpid, and so they remain until the bright sun and balmy air of spring wake them from their long sleep, and call them again among the flowers. At once they separate, and each, widow though she be, makes a home and founds a colony of her own. She finds a spot which suits her, and begins to dig a path in the ground. She picks out the grains of dirt with her strong jaws, passes them from one pair of legs to the next, under herself, and finally kicks them as far behind her as she can. When her passage is deep enough, a few inches or even some feet long, she ends it in a rounded cavern, which she lines with soft leaves. Some-

times she borrows the burrow of the field
mouse, and quite often the field mouse comes
and reclaims his own.   Indeed, he is not careful to prove ownership, particularly if the
chamber is well filled with honey and young
brood.

When the room is done, she builds brood-
cells, taking the wax from herself, like the hive
Bees.   Her comb is not built in the mar-
velously regular style of the hive Bees.
She makes an egg-shaped cell of dirty
wax, shaped like an earthen jar.   This she
places on its end, mouth upwards.   Then she
sets another beside it, and so gathers an irregular mass of cells, some standing on the
ground, some fastened to the walls of others.
Some are filled with honey; others receive
eggs.   If more than one tier of cells is found,
the second and third will be placed above the
first, and will be supported by waxen pillars.
Besides these cells, others are built by themselves about the room.   These are filled with
honey.   The honey jars are never sealed up,

for they are not filled for winter supply, but for daily use.

In about fifteen days from the laying of eggs, the labors of the mother Bee, who has hitherto toiled alone, are rewarded by the appearance of workers. The young Bees make more comb, and fill the cells with honey and farina. They line the roof and walls of the nest with a coating of wax, to keep the earth in place, and to prevent the rain from soaking through. When the new cells are ready, the mother lays a new supply of eggs. She must protect the new laid eggs from the workers, who would eat them if not driven away. At times she gets angry at some who persist in their efforts to get the eggs, and chases them out of the nest; but her wrath has defeated her prudence — the others take advantage of her absence, and steal her treasure.

If she can guard the eggs for a few hours, the danger ceases. In four or five days they are hatched, and as soon as the grubs are grown, each spins a cocoon for himself. Several eggs are placed in one cell. As the grubs grow, the cell becomes too small, and the pressure tears it open. The Bees patch up the rent. Presently it tears again, and again it is patched. Thus in a little time it becomes four or five times as large as it was at first. The patch work is not fitted neatly, like the wax work of the honey Bees, and produces the rough, clumsy cells found in these nests. The males are more useful than the drones in the Bee hive; for though they do not gather food, they provide their share of wax. The other Bees do not kill them in autumn, but all perish together when the frosts come.

These underground cities frequently contain quite a dense population. In one nest were counted 157 males, 56 females, and 180 workers, making a total census of 343. These numbers seem small compared with the 20,000 to 40,000 honey Bees in a hive, but if we remember that the Humble Bees are much the largest, that the comb is large and very irregular, we find that so many require a large space; and we must not forget that they usually dig the place for themselves in the earth.

Their honey is very sweet, but is apt to give headache. The wax is not clear like ordinary beeswax, and will not melt as well. Each species makes a cell peculiar to itself, either in position or shape.

Huber, while studying the habits of these Bees, placed several under a glass, with a piece of brood-comb. He took away all their wax and honey, and gave them farina only. The comb did not rest fairly on the table, and when the bees climbed upon it, to make it warm enough to hatch the eggs, it rocked to and fro. This motion annoyed them very much, but they had no wax, and could not make props to keep the comb in place. A few of the Bees then rested the hooks of their hind feet upon the comb, and braced the middle and fore feet upon the table. In this way they propped the mass on every side, and kept it steady. They remained in this position until relieved by others, taking turns together for two or three days. Then Huber gave them some wax, which they at once wrought into pillars, beneath the comb. But in a few days the wax became dry and gave way, and the Bees had to support the comb as before.

One variety of Humble Bee does not dig a chamber in the ground, but fills up a crevice in a heap of stones, and for this has been called the Lapidary Bee, *Bombus lapidarius.*

Another is the Carder Bee, *B. muscorum.* This Bee makes a nest in some hollow upon the surface of the ground. It consists of a roof of moss, lined and bound together with moss. It has an entrance at the bottom, which is also covered with an arch, and the whole affair is shaped not unlike the huts which the Esquimaux build of snow. The manner in which the Carder Bees prepare the moss for their nest is quite curious. When several have found a supply which suits them, they form a line from the nest to the moss. The foremost Bee takes a bunch of moss and combs it with her jaws and fore feet until the fibres all lie straight in a bundle ·beneath her. She then pushes it behind her, and at once proceeds to make another bundle. A second Bee takes the first bundle, combs it again, and kicks it back to a third, and so it is passed on from one to another, along the whole line to the last Bee, which puts it in its place on the roof of the house. This domed roof is made from four to six inches high.

Certain kinds of Bees have been called False Humble Bees, or Cuckoo Bees, *Apathus.* They are like the true Humble Bees in size and shape, but they lack the brush-lined cavities in the thighs for carrying pollen. These Bees do not build any house, do not make cells, or store honey, or care for their young. They are rovers, who take care of number one, and lay their eggs in the nests of other Bees. The larvæ which hatch from these eggs are stronger than the rightful occupants of the cells, and eat up all the food. So the hard working Humble Bee has built her cell for an intruder, and continues to care for it as if it were the

true heir, which it has starved out. Such
things do not happen among mankind alone.

Among the solitary Bees several trades are
represented. Their labors all tend to the same
result — shelter and food for their young, while
some work in wood like carpenters; others,
like masons, build houses of mortar; others
excavate the ground as miners; others find
cavities, which they line with leaves, like up-
holsterers.

The Carpenter Bee begins her work in early
spring. She chooses a bit of wood which
suits her, usually the dead branch of a tree,
or a weather beaten board, and in this she
bores a hole about an inch and a half long
and large enough to turn round in, which
usually opens upon the under side of the
branch or board, so that the rain may not come
in. After boring directly in as far as she
chooses, she turns and works several inches
along the grain of the wood. All her chips
she takes out and stores carefully in some place

where they will not be blown away by the wind.

When she has bored as deep as she chooses, she begins to fill up the hole again. She puts a little heap of pollen in the bottom, and lays an egg. Then she goes to her store of chips and gets material for a floor above the egg. She fastens the chips in a ring about the wall, with glue from her mouth. Within this ring she makes a second, then a third, until the partition is complete. On this floor she places another pile of pollen, and an egg; and thus she continues until the hole is full. When the egg hatches, the grub finds a supply of food; in a few days it has grown to its full size, and changes to a chrysalis, placing its head downwards. In this way the perfect Bee, as it gnaws its way out of the wood, is prevented from interfering with its younger brothers and sisters which are not yet quite ready to meet the responsibilities of society. English writers describe the Carpenter Bees as living in South America and Africa; they may be found in various parts of the United States.

A variety of wood-boring Bee chooses the stem of the willow tree for its home. When its tunnel is finished, it flies away to a rose bush, alights upon a leaf, and cuts out a round piece, about as large as a half dime. Many persons seeing the round spaces left, charge the mischief to the caterpillars. The Bee stands upon the piece which she cuts off, and as it falls she flies back to her nest with it in her jaws. She bends it into a cup shape, and stuffs it down to the very bottom of the hole. When the cell is suitably lined, she puts in some pollen and an egg, and covers it with another bit of leaf, which is the floor of a second cell. When the leaves are dry and stiff, they are so compact that the whole may be taken out together, and then separated into sections, like a row of thimbles thrust into each other. One variety of the Upholsterer Bee uses the scarlet leaves of the poppy for the silken lining of its cradle.

When a boy, the writer was somewhat frightened by a bee which came into his bedroom. The alarm was soon changed to curiosity, when the Bee was seen to examine an old inkstand, which had several holes in it for holding pens. The Bee would enter one of these holes, remain an instant, fly away out of the window, and presently come back to the same place again. So she buzzed about all that day and the next, and by the end of the second day she had filled up all the holes in the inkstand, and plastered them over neatly with mortar. She explored the central place, where the ink should be placed, but although it was dry, it did not suit her, and she departed. The holes were found to be divided into cells by partitions of mortar, and in each cell was a grub which would have become a Bee.

Other Mason Bees build a mass of cells, placed side by side, in a lump, which they stick against the side of a wall, or in a corner. They love to work in the dark attic of a house, where they are undisturbed, finding entrance through some crevice or knot-hole.

They frequently fill the hollow stems of old raspberry vines, and the smaller kinds fill straws or nail holes. In fact, they occupy all sorts of odd and queer places, even filling up the scrolls of a snail shell.

There is no better sport for a boy than the watching of one of the working insects in a quiet afternoon among the summer holidays. Unlike the birds, they do not mind the presence of a visitor, and go right on with their work. An ant hill, a Bee hive, a solitary Bee, a spider spinning his web, or a hornet building his paper mansion on the other side of the window pane, will pay for many an hour's silent observation. And the quiet boy, with watchful eyes, will find many chances of seeing them, which he least expected.

THE GREAT MYGALE.   *Mygale Cancerides*

About Spiders.

———

Articulata.— Insecta.

Order — *Arachnida.* Spider-family.

———

URIOUS and beautiful forms are found in every department of the insect world. In all its infinite variety there are none which do not pay for careful, watchful study. We have described two great tribes of workers. Each is busy, one not more than the other. " Go to the ant, thou sluggard," says the wise man, " consider her ways and be wise."

So doth the little busy bee
   Improve each shining hour;
And gathers honey all the day,
   From every opening flower.

We come now to the family of spiders. They are workers, too, in their way, but their labors are devised only to carry on their great business of preying upon other insects. They are carnivorous insects; made to live upon flesh, just as the animals of the cat tribe live upon other animals, and as the hawks prey upon other birds. They serve a very important purpose in the insect world, for they help to keep other tribes, which would increase too rapidly, in their proper proportion.

Many people have a natural dislike to a Spider. They are known to bite, that is, to sting — flies, at least — and there is a kind of fear that they may sting men or children. They seem to be very crafty, and then they run so fast, and in such unexpected ways, that young ladies think it quite proper to scream, or run, if a Spider happens to come her way. Then the

housekeepers hate them because they spin webs
in the corners; the webs gather dust, and the
room is untidy. The offending webs are swept
down, but the Spiders are diligent, and in a few
hours replace the webs. So the housewives
search diligently, and without mercy put the
persevering insects to death. It may be that
perseverance, as an abstract quality, is not as
valuable as some people think. Perseverance
in a good cause, to attain a desirable object, is
very commendable, but perseverance in an evil
way only makes the evil worse. We are apt
to think that ways which are not in harmony
with our ways are wrong, and so the housewife
very much dislikes the perseverance of the
Spider.

Goldsmith writes of a Spider which he
watched. It was three days making its web;
then another Spider came, and in the battle
which the two had for the web they nearly
ruined it. Three days more were spent in
repairing damages. When the web was complete again, a wasp was caught in it, and as the

Spider did not dare engage so powerful an enemy, it cut the bands and let the wasp go. But
the web was so torn that the insect thought
it easier to make a new one, than to repair
the old. This new web Goldsmith destroyed,
and the Spider made another. Again he destroyed the work, but the poor creature could
spin no more. It had spun four entire webs,
besides making repairs enough to complete another, and had worked nearly fifteen days. Its
only resource for a living was to drive another
Spider from its web, and take possession.

In shape and structure the Spiders are all
similar, but unlike most other insects. A wasp,
a bee, or an ant, has three distinct parts — a
head, a body or thorax, and a belly or abdomen ; and these three parts are connected by
slender cords or tubes. The Spider's head
and body seem to have been soldered into one
piece, as if a man's head were set firmly upon
his shoulders. Naturalists call this the *cephalo-
thorax*, or head-chest. Its body, as well as the
eight legs which are joined to it, is covered with

plate armor of strong scales. The fore part
has two branches, which might be called arms,
each furnished at the end with a curved sting,
shaped like the claw of a cat. Each claw has
a tiny opening near the point, through which
poison passes into the wound which it gives.
When a fly is caught in its toils, the Spider
runs to it, and strikes with these arms, inflict-
ing wounds with its poisoned dagger-claws. In
different parts of the head the Spider has several eyes, generally eight, but sometimes only
six, and these eyes are arranged differently in
different species. The number seems to make
up for their want of motion.

The hind part of the Spider is covered with
fine supple skin, and clothed with hair. Near
the end are four, five, or six, little swollen spots
or spinners. Each of these has a multitude of
little tubes, so many that the microscope has
shown a thousand in a space no bigger than a
pin's point. Out of these tubes comes the material of the Spider's web. At a little distance,
the threads from all these tubes of one spin-

ner join, and then the strands from all the
spinners are joined together.   Thus the thin
spider-line which one can barely see, as it glit-
ters with moisture in the sunshine, and in many
positions can not see at all, is made of four or
six strands, each strand composed of more than
a thousand threadlets.   This wonderful cable is
strong enough to support the Spider herself.
She often stops spinning in mid-air, turns back
and climbs up the same cord to the place
whence she let herself fall.

The spinning of the Garden Spider is probably not more curious than that of any other, but
it is rather more easily observed.   Sometimes
one begins her web on the outside of a window,
and is easily watched from within.   She begins
by pressing the spinners against the wood of
the window frame; a little of the gum exudes,
and fastens one end of the line.   She runs
along, giving out line as she goes, until she finds
a good place to fasten at, where she presses herself against the wall, making the other end secure.   She first stretches a few lines about the

space which the web is to fill, forming a triangle, or a four-sided figure. She then draws a line across the middle of this space. All these lines she makes very strong, doubling some of them several times. If any of them seems to become slack, she fastens a line near one end, and pulls it aside, until the main line is taut. Now she goes to the middle of the cross line, fastens a line there, and then runs back to the margin and fastens it an inch or so from the end of the cross line. She goes to the middle, and stretches another line in another direction, and then another, as if she were putting in the spokes of a wheel. While doing this she does not put in the rays or spokes on one side first, but draws her lines in opposite directions, keeping the strains all the time even. When she is about the first part of the work, running the marginal lines, and placing the first few spokes, she works slowly, stopping now and then to plan ; but as the web progresses, she seems to have solved her problem to her satisfaction, and hurries on the work.

Presently the rays are all set. Then she goes to the centre, and lays down a spiral line, fastening it to every spoke, and drawing it round and round, at even distances, in ever widening circles, until she comes to the outside. The main lines and rays are made stout and firm. The spiral lines are very elastic, and may be drawn far out of place without breaking. The garden Spider finishes her web in a few hours. She works as well by night as by day; in the dark as in the light.

When her web is done, she hangs herself in the middle of it, with her head downwards, waiting until some insect becomes entangled in her snare. When she feels the web move, she rushes to the spot. If the game be small, she thrusts in her dagger, and kills it at once. If it be large, and there is danger that its struggles will tear the web, she at once winds it round and round with cords, which she spins as she goes. She ties it, wing and foot, until its struggles can do no harm; then she gives the fatal blow, and eats the victim at her leisure. If the insect is so large that she can not manage it, she cuts away the threads as quick as possible, and lets it go, before it has torn her web in pieces.

A writer for the " Atlantic Monthly," a surgeon in the United States Army, gives an interesting account of the spinning of a kind of Spider, *Nephila plumipes*, which he found on one of the sea islands near Charleston. These Spiders were quite large. The females were from an inch to an inch and a quarter long; the males were only about one fourth of an inch long, and about one hundred and twenty would have weighed as much as one of their buxom wives. Accident showed him that he could reel the silk from the living Spider. He therefore gathered as many as he could find, and brought them north to experiment with. When ready to spin, he fastened each in a little frame of cardboard, which would hold the insect without hurting it. Then he reeled the silk upon a suitable reel. From one he wound about one thousand yards, and from another over *two miles* of silk. A single thread sustained a weight of fifty-four grains.

The silk from the same Spider was of different colors and qualities. At the same instant he wound from one insect one thread golden yellow, and another bright silver white. If the two ran together, they made one light yellow thread. The white silk, when dry, was firm and unyielding, suitable for the rays of a Spider's web. The yellow was very elastic, like that used for the spiral rings which bind the rays together. There was also a pale blue silk which seemed to be used to tie up an insect after it was caught in the web. Enough silk was reeled to be woven in a loom, upon a warp of black silk, so as to make a bit of ribbon two inches wide, showing that it was real silk.

The House Spider usually puts her web in some corner. She runs out as far as she intends to spread the web, fastens a thread to the wood, then goes back to the corner and out on the other side, until she comes opposite the place where she first made the thread fast, and there fixes the other end. Then she places a second and a third thread beside the first, for these make the foundation of her whole work. From these she draws other lines to the angle, and then she works back and forth over the whole, until the piece of gauze is done. She then stretches a great number of threads from side to side above her web, crossing them every way. These lines are arranged not unlike the tackling of a ship, and often reach two or three feet high. The flies passing through the space become entangled, fall upon the web below, and are caught. Besides all this, she makes a round funnel, for a hiding place, below the web, in the corner, or behind some piece of furniture. Here she waits and watches, out of sight. If the least touch disturbs the web, she feels it, for the rays from every part pass down into this funnel, and she rushes forth to learn the cause.

A Spider of Jamaica is called the Trap-door Spider. This insect digs a burrow in the ground, and lines it first with coarse, rough web, which seems more like the paper of the wasp's nest than the silk of the Spider. The inner lining is smooth and soft, and may be drawn out of the other, without injuring either. The tube is placed where the surface of the ground is a little sloping, and the mouth is covered with a door, made like the lining of the tube. This door is fastened by a hinge at the upper edge, in such a position that it falls into place by its own weight. The outside is covered with earth, which perfectly conceals the nest. A stranger may well be startled at seeing a hole open in the ground at his feet, and a large Spider peep out to observe what is going on. One of these Spiders dug its tube in cultivated ground. After it was made, the earth was heaped over it about three inches; the Spider finished out its tube, and made a second door at the new surface.

This Spider is about an inch and a half long. It leaves its burrow at night and hunts for its prey. If any one attempts to raise the trap, it hooks its hind legs into the door, and its fore

legs into the side of the tube, and holds on with
all its might. It will suffer its nest to be dug
out of the ground and carried away without
leaving it; in this way they have been caught
and put where they could be watched. Other
species which make their home thus are found
in Australia and elsewhere.

In Surinam, and on the Amazon river, Spiders are found of the genus *Mygale*, which
destroy birds. When this was first reported,
it was not believed, but the Spiders have been
caught in the very act. When we consider the
size which they attain, the wonder ceases. One
is described as two inches in length of body,
and more than seven inches in expanse of legs.
It was covered with coarse red and gray hairs.
Some of these huge Spiders make a dense web;
one digs a burrow two feet deep, and lines it
with silk. When the children catch one of
these fellows, they tie a string about its waist,
and lead it along like a dog. The Mygale
sheds its hairs easily and they pierce the skin

of one who handles it, causing painful irritation.

The name Tarantula is given to several large
Spiders that live in the ground and hunt for
prey. The Italians have a belief that one kind
will cause a disease which can be cured only by
dancing a long while to peculiar music. The
sting really makes but a slight wound.

One member of this family lives in the water.
Still it lives by breathing air, and therefore it
takes a supply along with it down under the
water into its nest.   Like all the other Spiders,
this makes its nest of silk ; it is generally about
as large as an acorn, egg-shaped, and open
below.   This cell is filled with air; and if the
Spider be kept in a glass vessel, it may be seen
in its cell, resting in Spider fashion, with its
head downward.   Where the air came from
was, for a long time, the question.   Some
thought it was the oxygen which was formed
by the water plants.

A few years since, Mr. Bell saw some of
these Spiders spin their webs, and fill them

with air. When one had made her web, she went to the surface, grasped a bubble of air, descended quickly to her nest, and thrust the air in. Then she came up for more, and after twelve or fourteen journeys she had laid in her supply. When enough had been collected, the Spider crept in and settled herself to rest in her transparent cell.

" The manner in which the animal possesses itself of the bubble is very curious. It ascends to the surface slowly, assisted by a thread attached to a leaf below and to one at the surface. As soon as it comes near the surface, it turns the extremity of the abdomen upwards, and exposes a portion of the body to the air for an instant, then with a jerk it snatches, as it were, a bubble of air, which is attached not only to the hairs which cover the abdomen, but is held on by the two hinder legs, which are crossed at an acute angle near the extremity, this crossing of the legs taking place the instant the bubble is seized. The little creature then descends more rapidly and regains its

cell, always by the same route, turns the abdomen within it, and leaves the bubble."

The water Spiders feed on the insects which
swarm in the water, eating their prey in their
homes.

Another aquatic Spider builds a raft. It
gathers together a mass of dry leaves and similar things which will float, and fastens it with
silk threads. On this raft it sits, floating wher-
ever the winds and waters carry it. When the
water insects come to the top, it seizes them
before they can escape. Others fly over the surface for their prey, and fall into the jaws of this
Spider-wolf. It is quite large, and very beautifully colored and marked.

At certain seasons of the year, large quantities of gossamer threads are seen floating in the
air. They fall upon the grass and streak it
with fine lines. They gather on the trees.
The steamboat, plowing up the long lanes of
water through forest and prairie, gathers
streamers and pennons of gossamer on every
pole, and the rough helmsman frets as the films

catch upon his eyebrows, and dim his sight. All this is made by Spiders. They climb to the tops of trees, and pushing the gossamer out at their spinners, let it float upon the air until its buoyancy is enough to carry them away. Balloonists have found these Spiders floating in the air above their cars.

Says Gilbert White: "Every day, in fine autumnal weather, do I see these Spiders shooting out their web and mounting aloft. They will go off from your finger if you will take them into your hand; last summer one alighted on my book, as I was reading in the parlor, and running to the top of the page and shooting out a web, took a departure from thence. But what I most wondered at was, that it went off with considerable swiftness, in a place where no air was stirring; and I am sure I did not assist it with my breath; so that these little crawlers seem to have, while mounting, some locomotive power, without the use of wings, and move faster in the air than the air itself."

There are spiders which lie concealed in a

rolled up leaf, and seize any insect which comes in the way. Others lurk in the cup of a flower, and eat the fly that comes for honey. Some hunting Spiders leap upon their prey like tigers, and have a way of jumping sideways. They steal upon their game as a cat steals upon a bird. If the fly moves, the Spider moves too — backwards, forwards, or sideways — until the two seem to be moved by one unseen spirit. If the fly takes wing and alights behind the Spider, it turns about with the swiftness of thought, too quick for the eye to follow. When its movements have brought it within reach of its victim, its leap is sudden and deadly as lightning.

The Spider is very watchful over its young. Most species do not lay eggs until two years old. Then the female prepares a cocoon of silk, very thick and strong, in which she places from fifty to a hundred salmon-colored eggs. This sack is often made of two dish-shaped pieces, fastened together at the edges. Sometimes it is hidden in the crevice of a wall, or under the edge of a loose board. In this case it is securely fastened by a net-work thrown over and about it. It is often carried about by the mother, attached beneath the abdomen, or held in the jaws as a cat carries her kitten.

If any attempt is made to carry away this treasure, which the mother always watches over, she resists it to the utmost. When taken from her, she becomes listless, as if stupefied; if restored, she seizes it eagerly, and runs away with it to a safer place. When the young are hatched they remain in the cocoon until, at the proper time, the mother bites it open and sets them free. Even then they do not leave her, but remain, like a brood of chickens, under her care. She often takes them upon her back; she provides food for them, and leads them about until they have age and strength to shift for themselves.

The gentleman who obtained the silk spinners from Charleston harbor, procured a large number of these egg sacks, and in a short time had a brood of about *two hundred thousand.* One

bright June day he left them on a tray in the
sun, and on his return found his brood — *baked.*

A supply of Spiders, which he kept in little
paper boxes, furnished a fresh harvest of eggs,
from which about seven thousand were hatched.
They appeared in about a month after the eggs
were laid.   For a long time they seemed
to eat nothing;   then they shed their skins,
and began to grow.   As they grew, their
numbers diminished, and it began to be evident
that they were eating each other.   Shut up in
the sacks they had nothing else to eat, and the
weaker ones were a prey to the stronger.   They
were then placed in inverted glass jars, with
wet sponges in the mouths, and were fed with
flies, bugs, and afterwards with such flesh as
bits of chicken's liver.   Some of the first family brought north seemed to go into a decline
and die, for no cause which their keeper could
understand.   He tried various expedients with
them, but nothing did any good.   At last he
thought of giving them water, although he had
never known that Spiders drank water.   A

drop was given on the tip of a camel's hair pencil, and was eagerly seized. All the Spiders drank, some taking several drops. Besides water to drink they required some moisture in the air. They became quite tame; would eat and drink from a bit of stick, or a pin, and when stroked gently, would raise up the back like a cat, or put up a foot to push away the finger.

As was said before, the Spider is a type of industry and perseverance, no less than the ant or the bee. The Scottish farmers love to tell that King Robert Bruce once learned a lesson of endurance from a Spider. While wandering on the wild hills of Arran, he passed a night within a poor, deserted cottage. He threw himself down upon a heap of straw, and lay, with his hands under his head, unable to sleep, but gazing up at the rafters of the hut, festooned with cobwebs. From long and dreamy thoughts about his hopeless condition, and the many evils which he had met, he was roused to notice the efforts of a poor Spider, which had begun its work with the first gray light of morning. The insect was trying to swing by its thread from one rafter to another, but it constantly failed, each time swinging back to the point from which it sprang. Twelve times the little creature made the attempt, and twelve times it failed. Without delay it tried again, and the rafter was gained. " I accept the lesson," said Bruce, springing to his feet; " I shall again venture my life to win the battle for my country." And the victory was won.

WINGED ANT-LION.—*Myrmeleo libeiluloides.*

ABOUT DRAGON-FLIES.

ARTICULATA.— INSECTA.

ORDER — *Neuroptera.* Net-winged.

FAMILY — *Libellulidæ.*

A LONG, slender insect, with large head, swollen on either side by a huge eye, flying with four broad, gauzy wings, is a frequent mid-summer visitor. He and his mates range up and down in the air, pausing here a moment, then darting away in the most unexpected manner. He comes into the house with a great buzz, and makes vain attempts to fly back to free air through the window pane. He seems to have no particular business, except flying about, buzzing, and bumping his head. The children call him a Darning-needle, because his body is straight and slender; and as its long and flexible tail twists about more than seems pleasant, they are afraid of it; they believe it can sting, and some call it a Horse-stinger. But the creature has no' sting, and can do no harm to man or beast. In the insect world he well deserves his name, Dragon-fly, for he devours multitudes of other insects. When dancing in the sunshine, or in the twilight shadows, he is busy catching gnats, or sweeping up other minute specks which fly in the air. He is not content, even, with such small game, but is the eagle among insects, pouncing upon unwary butterflies, which he drags to some bush to devour at his leisure.

The water is his birth-place. The eggs, like a bunch of grapes, sink to the bottom and hatch out six-footed larvæ, with dusky brown skins. Like many other grubs, when these youngsters grow too large for their clothes, they split them open, throw them away, and soon appear in a new and larger suit. When full grown, a pair of scales appears on the back, which is a mere suggestion of wings. The head is then armed with a long, jointed trunk, fitted at the end with a pair of strong hooks. While at rest, this trunk lies folded over the face, like a mask; if any prey passes by, the trunk leaps forth, and the hooks grapple the unwary victim.

The Dragon-fly lives as larva and pupa, two years. When ready to come out into the world, it climbs to the top of some water plant, into the sunshine. The eyes show when the change is coming. Instead of dark, dull places where eyes might be, they become clear and bright, and the real eye shines through the mask. If one can be found at this crisis, and fastened where the change can be seen, it will yield much amusement.

First a rent comes in the skin along the back, to the face; here another rent opens crosswise, over the eyes. Now that he has burst his case, he carefully picks out his legs, and then hangs his head down, motionless, as if dead.   He has only hung his moist legs out to dry.   Presently he lifts himself again, grasps the case with his feet, and slowly draws out his long tail, and wet, sodden wings.   But the tail has not its full length, and the wings are folded. He rests awhile; the tail expands, the wings unfold, and as they harden, glisten like sheets of mica.   While in this wet condition, the Dragon-fly is careful not to touch them, even with its body; for a wrong twist now would make a deformity for ever.   The change may be passed in a quarter of an hour, or may take several hours, according to the clearness of the air.   When the wings are fully spread and hardened, and the bright colors of the mailed body are fully set, he leaves his twig and begins his long journey through the air.   Like a newly commissioned Alabama, armed and supplied for a long cruise upon the high seas, he sets forth, a piratical rover, to capture, plunder, and destroy.

While living in the water, this creature has

a way of moving about peculiar to itself. If
seen at the bottom of clear water, it seems to
move merely because it wills to move, with
nothing like walking or swimming — it goes.
But if a few grains of sand be near, they seem
to will to go backward, at the same time. Put
one of the larvæ into water colored with milk
or indigo, and then suddenly change him into
clear water, and the motion will be explained.
He will be seen to spirt a stream of colored
fluid into the clear water, and it will be found
that he has in his abdomen a set of force
pumps. These fill slowly from the fluid in
which the larva floats, and then drive out the
water backwards, while the same force which
ejects the water, pushes the insect forwards.
Some English ship-builders propose to drive
steamships by this plan, which it may be they
borrowed from this very insect. They take
water through the bottom of the ship, and then
drive it out astern by powerful steam pumps.
In this way they expect to force the vessel rapidly through the water.

We nave mentioned the large globes of eyes on either side of the Dragon-fly's head.   Under a small lens these eyes seem to be covered with fine net-work.   A magnifier of larger power shows that the surface is composed of regular, six-sided faces, so that it resembles a minute crystal honey comb.   Farther examination shows that each eye contains more than 12,000 of these lenses, and that what we call the eye is only a bundle of eyes.

Opticians grind a multitude of flat faces on a rounded bit of glass, which they set in a tube. Any thing seen through this tube seems multiplied as many times as there are faces on the glass; the image is very pretty, but very much confused.   We need not suppose, however, that the Dragon-fly is puzzled by his compound eye, or that he sees more than one image.   Although we have two eyes, we do not see double.   The nerves which carry word to the brain that the eyes see something, meet just behind the eyes, and perhaps, for this reason, report but one object.   If two eyes thus unite their results, so that we do not see double, in the same way 25,000 eyes in one head may combine all their results. The fact that we see so many images in the multiplying glass will not trouble us if we remember that our own eye is behind the glass, instead of a bundle of nerves, and there is no way of gathering all the images into one.

There are many species of Dragon-flies, strong of wing, and beautifully colored with bright blue, green, scarlet, glossy black, or transparent white. The body is often of one hue, while the wings are barred or spotted with others. Often the male and female of the same species are variously marked. These bright colors always vanish when the animal dies; in a few days the most brilliant specimens will have faded to a blackish brown. The only way to preserve them is to remove the interior substance, and fill the space with paint of the proper color, and this method does not repay the time and labor spent.

One tribe belonging to this family are called

Scorpion-flies. The rings near the end of the tail are quite slender, and move easily in any direction. The last ring is stout and thick, and bears a strong pair of forceps. When the fly is at rest, the tail is curved over its back like that of a puppy, but when alarmed it flourishes the tail in a very alarming style, the forceps snapping as if something serious would happen if there were a chance.

Some other members of the order *Neuroptera,* or nerve-winged insects, are worthy of notice.

The large, prominent eyes of the Lace-wings, or Golden-eyes, glow with changeful flames of gold and ruby, as if on fire. These insects are small, but their brilliancy and their broad wings make them quite conspicuous. The larva of the lace-wing is very voracious. It is particularly fond of the plant lice, and therefore is quite useful. A single one will clear a densely crowded twig in a short time. It will, however, turn and eat the eggs in which its brothers are ready to hatch, if it can reach them. To prevent this, the instinct of the

mother makes her spin a slender thread, like a
bit of bristle, about a third of an inch long;
the lower end of this thread she glues fast to
a twig, and on the upper end she leaves an egg
about the size and shape of the letter o. So
she places a dozen in a group, which is easily
mistaken for a patch of moss. For a long time
these were really supposed to be a variety of
moss, nobody suspecting that they were the
eggs of an insect. When the first hatches, he
falls down upon the twig. He reaches up to
breakfast on another egg, but he can not climb
the slender waving stalk, so he creeps away,
and finds his meal elsewhere.

A somewhat celebrated insect of this family
is the Ant-lion. In its perfect state it much
resembles the Dragon-fly, but the wings are
broader and softer. ·It is most remarkable
when a larva. Then it resembles a flattened
maggot, with long legs and large jaws; but the
legs are of little use for walking, as it moves
mostly by means of its abdomen. It is very
slow, and yet very voracious, living on insects

much quicker than itself, which it catches alive.
As it can not take them in open chase, it sets
an ambush by digging a pit, and lying concealed at the bottom. In this work it begins
at the outside. It presses its body down into
the sand, and then backs round in a circle,
plowing the earth and throwing it outward.
So it goes round and round, drawing one furrow after another until it comes to the middle.
This plowing is repeated several times, as long
as it will turn the earth outward. Then it
begins to dig. It goes to the middle, and flings
the sand out with its head, and smoothes the
sides of the pit, down to the centre, into a regular funnel. If it finds small stones, it jerks
them, one by one, over the wall. If too large
for that, it takes them on its back and carries
them up the slope, and tumbles them over the
edge. Sometimes, after toilsomely tugging until
a stone is nearly at the top, the pebble topples
off and rolls to the bottom again, plowing a furrow as it goes down. The Ant-lion tries again,
pushing the load up the same furrow; he

works on until the stone is removed, or until repeated failure satisfies him that he is not equal to the task. Then he leaves the unfinished pit, and digs another.

When finished, the pit is about two inches deep, and three inches in diameter. The Ant-lion lies at the bottom, only his jaws being in sight. When an ant, journeying that way, looks over the edge, the loose sand under its feet begins to slide, and lets it down into the pit. It struggles to regain the top, but that only hastens its fall, and down it goes into the jaws of the hungry monster which waits for it at the bottom. If the ant succeeds in climbing up, and is likely to get out of danger, the Ant-lion shovels sand upon its head, and flings it after the escaping insect. Overwhelmed by this storm the ant is borne to the bottom. When the juices are sucked out of him, the empty skin is tossed over the mound, and the pit is put in order for the next unfortunate.

Thus the Ant-lion lives for about two years. Then it wraps itself in a covering made of sand glued together, and bound by a kind of silk which it spins.   In about three weeks it emerges in its perfect form.

Another of the *Neuroptera* is the May-fly, or Ephemera.   The early days of summer bring vast swarms of them, which vanish as suddenly as they come; often a single day is sufficient for the entire round of their perfect life.   Hence the name Ephemera—"(lasting) for a day."   It is, however, a mistake to suppose that a day is enough for the entire life of the insect from the egg to the grave.   On the contrary, two years are passed in the water before the winged form is assumed.   Like other creatures that flit a few brief days about watering places—although it does not carry a Saratoga trunk full of finery —it can not do without a change of dress.   So, after dancing its set in one costume, it retires to its chamber—a twig—kicks off its garment, and appears in another, bright and new, with larger wings, broader plumes, and longer train.

In both dresses, the May-fly is very eagerly taken by fish, and adroit anglers use them, or

SECTION OF TERMITES' NEST.

imitate them, when they would bring wary old trout from their deepest hiding places. Very much alike — Newport belles, and Newport May-flies!

TERMITES.

THE remarkable insects known as Termites, or White Ants, though commonly called ants, are not classed with that order, but among the *Neuroptera*, on account of the structure of their wings in their perfect stage. Like the ants, the Termites live in societies, which become immensely large. They build for themselves huge cities, great mounds, conical like sugar-loaves, sometimes twenty feet high, and more than a hundred feet in circuit. They make these of clay, and so solid and strong, that the wild cattle climb on them without breaking through. Within they are full of chambers and passages. There are apartments for the king and queen; nurseries for the young; garrisons of soldiers; dwellings for workmen, and storehouses for food. These edifices are said to surpass the dwellings of ants, bees, and beavers, as much as the architecture of Europeans excels the rude huts of Indians or Bushmen. Some species build in the ground, partly beneath and partly above the surface; others build on branches of trees, and often at a great height.

One of the best known species is the *Termes bellicosus* of Africa. In Senegal, and parts of Central Africa, their numerous clusters of hills resemble the huts in the native villages. The first hill which they make, in beginning a settlement, rises above the ground perhaps a foot. While this grows larger and higher, others spring up at a little distance, and still others, until a circle of small hills surrounds the larger one in the centre. These all keep on growing; presently they join each other, and the middle cone includes or covers up the outer ones. Mean while the inside works which were first made, are pulled down, and the materials are used for building the outer cones. They have no precise form, the only care being to make them firm and strong. Until they are six or eight feet high they are quite bare, but after that they increase more slowly, and grass often grows upon them. In the hot season, when the grass becomes dry, the whole resembles a large haystack.

The royal apartment, as the most important room of the house, is placed in the centre. It is shaped like half of an egg, cut lengthwise, and is at first about an inch long; it is afterwards enlarged to suit the increased size of the queen, until it is six or eight inches long, or even more. The openings through the walls and roof of this room are large enough to admit the workers which are in attendance, but the royal occupants can never pass out; they are life-long prisoners. A set of chambers about the royal cell contains the soldiers who protect, and the workers who serve the regal prisoners. These rooms are connected

together; they extend a foot or two all round
the central apartment. They are surrounded
by the nurseries and the storehouses. The
latter are built of clay, and filled with gums
and similar vegetable substances. The walls
and partitions of the nurseries are made of
woody fibre, cemented together by the saliva
of the insect. When the nest is small, they are
near the royal chamber. As the family grows,
and the attendants of the queen become more
numerous, the nurseries are moved farther
away. They are enclosed in clay chambers,
like the granaries, and the wooden partitions
and linings would seem to prevent too sudden
changes of temperature.

A large arched open space, two or three feet
high, is left under the central dome, with
arched passages on every side, which allow the
warm air to circulate freely, and keep the nurseries at a proper degree of heat. The shell
which forms the great dome is traversed by
large round or oval passages, several inches
wide. These ascend spirally, quite to the top,

opening into each other, and into the central dome at proper distances. Other passages of less size connect the larger ones, and others still lead far away under ground. Even if all the Termites within a hundred yards of a house were destroyed, those which live farther away would extend their galleries to the house, eat up the merchandise in it, and destroy every thing. If they can not go under ground in the way they wish, they make pipes along the surface, of the same material as their nest; they often carry these covered ways above ground over the deeper paths, and make frequent communications between them, so that they can escape by one, if they are attacked in the other.

Each village of Termites has a king and queen, an army of soldiers, and a population of laborers. There are about a hundred workers to one soldier; they are about a quarter of an inch long, very busy and very swift. The soldiers are half an inch long, and as large as fifteen of the workers. The winged or perfect insects are nearly an inch long, and their wings spread above two inches and a half. They are equal in bulk to two soldiers. The young Termites come out of the nest just after the first shower has opened the rainy season. The immense swarms fill the air as with dense white snow flakes. Every living thing seems to be their enemy. The ants fall upon them and eat them; birds come in flocks and pick them up; reptiles and ant-eaters devour them, and the black men gather them as the greatest delicacy. Not one pair in a hundred thousand escapes alive, but that pair will, by and by, produce a hundred thousand a day.

While the winged insects are flying, and being eaten, the workers are running about on the ground searching for them. If a pair is found, they are at once chosen king and queen, and their new subjects proceed to build them a house. They are shut up in a little clay chamber, with only one small entrance, too small to allow them to pass out. Presently the female begins to enlarge in a wonderful manner; and the house has to be enlarged to correspond. In time, it is thought about two years, she is about three-fourths of an inch wide and three inches long — specimens have been found of twice that length. Her body is now oblong, banded at intervals of half an inch with dark muscles. The transparent skin is of a fine cream color, through which the intestines, and the motion of the fluids, may be clearly seen. When she has reached this size, she produces about eighty thousand eggs a day. The attendant workers carry these away to the nurseries, where they are hatched, and the young provided with every thing needed, until they are old enough to shift for themselves.

When a person enters a piece of ground which is marked by many of the covered ways of these insects, he hears an alarm given by distinct hisses. After that he may search the paths for Termites in vain; they have escaped by the underground lines. The tunnels are made large enough for passing and repassing without trouble. They serve as shelter from

light and air, and particularly from the attacks
of other ants. When driven from these defen-
ces the ants pounce upon them, and carry
them to their own nests to feed their young
ones. If the defence is broken, the workers at once set about repairing it, and even
if three or four yards is destroyed, the place
will be rebuilt before the next morning. If
the gallery is often destroyed, it will be given
up and another made, unless it leads to some
favorite plunder. The main roads are made
deep under ground, going under the very foun-
dations of houses and stores, and come up
under the floors, or through the posts on which
the building rests. While some are boring the
posts through and through, and taking out all
their fibres, others climb the outside and enter
the roof. If they find thatch, which they seem
to like very well, they bring up clay and make
covered ways in and through the roof as long
as it will stand. Thus they carry away, bit by
bit, every sill, and post, and beam, floor, ceiling, and partition. The outside seems firm

and sound, but the whole will crumble at a touch. Sometimes they seem to know that a post sustains weight, and then they fill up the cavities which they make with clay, packing it in more solidly than man could. The posts are found filled with material as hard and compact as many kinds of building stone. They will eat the very mat on which a man sleeps. They carry away all the wood of his strong box, leaving a shell as thin as paper. They devour his books, his records, his correspondence. If a piece of furniture be left too long in one place, nothing will remain but the surface. A man may be rich to-day, and poor to-morrow from their ravages.

It is a difficult task to destroy them. Any thing which is washed with corrosive sublimate they respect, but this can not be applied to many things. If the house is broken into, the soldiers come to the breach to defend it. They may be destroyed, but they are not those which do the mischief. The workers are left, and the business of the village goes on just as before. The only plan which is at all sure is to continue pulling down the nest until the chamber of the queen is found, and she is destroyed. Then the others seem to be bewildered, lose courage, and finally abandon the nest.

About the year 1780, some bales of goods, brought from St. Domingo, were stored in La Rochelle, and in other French seaports, and thus the Termites were introduced. At La Rochelle they took possession of the arsenal, and of the prefect's house, invading rooms, offices, court, and garden. A stake driven, or a plank left, in the garden, was destroyed forthwith. One fine morning the records of the office were found ruined, though not the least trace of damage was seen on the outside. The Termites had mined the wood work, pierced the card-board, and eaten up parchments and papers, but had always scrupulously respected the upper leaf, and the edges of all the leaves. By chance a clerk raised one of the leaves which hid this ruin, and discovered the injury.

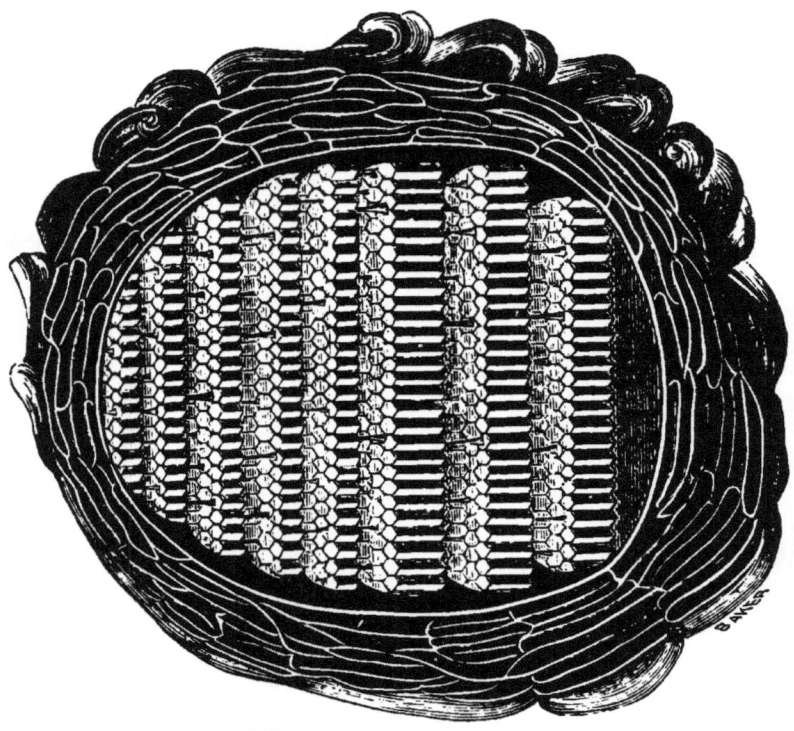

INSIDE.

A WASP'S NEST.

OUTSIDE.

ABOUT WASPS.

ARTICULATA — INSECTA.

ORDER — *Hymenoptera.* Membrane-winged

FAMILY — *Vespidæ.* Wasp-like.

WASPS attract attention, for two reasons. They have sharp, venomous stings, which they are ready to use on small provocation, and so make us afraid of them; and they build for themselves curious homes, which are well worth our study. Those that we are most familiar with, build with mud, or paper.

The paper makers usually choose some sheltered place, under a fence rail, in a bush, in a hollow tree, or under the projecting eaves of a house. As in the case of the humble bees, the mother of the family, single handed and alone, lays the foundation of the house, and makes preparation for rearing a family. She and a few like herself are the sole survivors of the thronged cities of last year. All the others perished at the coming of the frost which chilled her blood within her and kept her torpid till the warm south winds of spring awoke her from her long sleep.

When quite a little boy, the writer used to go away alone into a closet, to learn his lesson. The blinds at the only window in the room were always closed, giving barely light enough to read, when sitting on a stool beneath it. One spring day a Wasp came between the blind and the glass, and after much buzzing and much walking about, began to build. She first laid down, beneath the under edge of the upper sash, a patch of paper about a third of an inch in diameter; then, standing on this, she raised cup-shaped edges all about her,

increasing outward and downward, like the cup
of an acorn, and then drawing together a little,
until a little house was made just about the size
and shape of a white oak acorn, except that
she left a hole in the bottom where she might
go in and out.

Then she began again at the top, and
laid another cover of paper over the first,
just as far away as the length of her legs
made it easy for her to work. Now it was
clear that she made the first shell as a frame
or a scaffold on which she might stand to
make the second. She would fly away, and
after a few minutes come back, with nothing
that could be seen, either in her feet or in
her jaws. But she at once set to laying her
paper-stuff, which came out of her mouth, upon
the edge of the work she had made before. As
she laid the material she walked backward,
building and walking, until she had laid a
patch a little more than an eighth of an inch
wide and half or three-quarters of an inch long.
When laid, the pulp looked like wet brown

paper, which soon dried to an ashen gray, and still resembled coarse paper. As she laid the material, she occasionally went over it again, putting a little more here and there, in the thin places; generally the work was well done the first time.

So the work went on. The second paper shell was about as large as a pigeon's egg; then a third was made as large as a hen's egg; then another still larger. After a time the wasp seemed to go inside to get her material, and it appeared that she was taking down the first house, and putting the paper upon the outside. If so, she did not bring out pieces and patch them together as a carpenter, saving of work, would do, but she chewed the paper up, and made fresh pulp of it, just as the first was made. Of course the boy did not open the window, for he was too curious to see the work go on, and then he was afraid of the sting. How large the nest grew he never learned, for he soon after left the school, and saw no more of it. The Algebra and Latin which he learned that term were soon forgotten — he was really too young to study either, then — but he has not forgotten how the Wasp made her nest.

But he now knows pretty nearly what the Wasp did after his oversight of her ceased. She made the nest about as large as a goose egg, hanging with the broad end up, and with a hole as large as one's little finger at the bottom. She took out of the inside all but two or three thicknesses, and then she built paper combs in the vacancy. These paper combs were not made like the combs of the honey bees, standing upon edge, with the cells opening in the sides, but were hung to the top, with the cells opening downward. She made first a stout post or rope of paper, hanging from the centre of the room. To the end of this rope she fastened a floor, which she spread out flat and level until it nearly reached the sides of the room. Underneath this floor, which might quite as well be called a roof, or a ceiling, she made a number of cells, and laid an egg in each. It is not quite settled whether she builds the cells first, and then lays the roof over them, or whether she makes the roof first, and then places the cells under it; probably the two parts are made nearly at the same time.

As soon as the first eggs are hatched, the cares of the mother Wasp increase, for now she has a hungry family to feed. She must supply their wants, enlarge their cells, make more cells, lay more eggs, make additions to the house, and all together. Was ever poor human mother, left to bring up a family alone, more driven with work? In due time the older grubs are full grown, stop eating, and spin a silken cover over their cells. After a short season, having passed from grubs to pupæ, and then to perfect Wasps, they come forth. They take the heavy work upon themselves, and the toil goes merrily on. Day by day their numbers increase, and soon the mother Wasp has nothing to do but lay eggs in the cells which her children have made.

When the first tier of cells is full, another is made below it. Several pendant cords similar to the first, are fastened to various points of the tier above. Cells are hung upon them as before, and continually increased in number, until the several parts unite to form a second complete tier. The mouths are placed downwards, and the roof serves as a floor on which the Wasps walk when taking care of the young brood. As among the humble-bees, the first Wasps that come out are workers. The males and females are not seen until autumn. A large nest may contain seven or eight thousand cells, and each cell is occupied, on the average, by three tenants in succession. All the young grubs have to be fed; not with honey, as young bees are fed, but with animal food, usually flies. We can easily see that a good sized Wasp's nest, or vespiary, may be quite a serviceable thing about the house, if, in the end, the Wasps do not become the greater nuisance.

Mr. Wood says he has seen pigs, covered with flies, lying in the warm sunshine, and the Wasps pouncing upon them and carrying them off. It was a curious sight to watch the total indifference of the pigs, the busy clustering of the flies, which actually blackened the hide in some places, and then to see the Wasp just clear the wall, dart into the dark mass, and retreat again with a fly in its fatal grasp. On the average, one Wasp came every ten seconds, so that the pig-sty must have been a valuable store house for them.

The Wasps are hearty eaters, as well as their grubs. They prey upon other insects, sugar, meat, honey, and fruit. Indeed, they are particularly fond of ripe fruit, and always select the finest specimens, just when they are in their best condition, gnawing holes in them, and spoiling them for the table. Still it may be a question whether the good they do in destroying flies and young caterpillars does not more than pay for all the fruit they eat.

The nests of the paper-making Wasps usually vary from six to twelve inches in diameter. They sometimes become very much larger. A nest is preserved in a museum in Oxford, Eng-

land, which fills a glass case four feet high, by
two feet in width. It is turnip shaped, with a
large knob at the top by which it hangs. This
nest, when found, was about five inches in
diameter. It was taken into a house, and hung
near a window which gave the builders free
passage to the open air. There was no danger
in this, as the common Wasp has a much better
temper than the hive bee, and is by no means
as capricious in the use of his sting. Their captor was disposed to give them every means of
living, and supplied them daily with sugar and
beer. They consumed daily a pound of sugar
and a pint of beer. With plenty to eat they increased rapidly, and the nest grew as fast. In
the chamber above, two other nests had been
placed, and as those workmen were not fed, when
they found that their kinsmen below were
faring so sumptuously every day, they deserted
their own houses, and joined the colony on the
ground floor.

The Chartergus Wasp of Ceylon, another
paper maker, uses its nest as a permanent

home, the same family living in it from year to year. This home is enlarged in a way which keeps its shape, and allows farther increase without trouble. The walls are shaped like the sides of a cow bell. The tiers of cells extend from side to side, like the regular floors of a house. When the house is full, another set of cells is built beneath the lowest floor, the wall is lengthened down as far, and a new floor is made to shut up the bottom; so that the new house is the old one with a new story *under*. In fact, probably all the Wasps learned to build by reading Gulliver's Travels. The bells of this Wasp are usually about a foot long; one is described which was six feet long, and of corresponding width.

A South American Wasp has been called *Myrapetra*. It builds a nest of a dark, blackish brown substance, like *papier mâché*. The outside of the nest is thickly studded with projecting spikes or thorns. Their exact use is not known; some have thought that they are to protect the nest from wild beasts;

others suggest that they are meant to conceal the entrances. The tiers of cells are not flat, but shaped like inverted bowls; the dishes grow broader and flatter towards the bottom of the nest.

The other branch of the Wasp family includes the Mud-diggers, or Dirt-daubers. Up in the attic of any old house in our country, east or west, the children will often find, stuck on the walls and rafters, lumps of mud of various sizes and shapes. Some are as thick as one's finger, others as large as one's fist. If one of these shapeless lumps be opened carefully, it will be found to be a mass of cells, each lined with a thin coat of brittle, shelly substance. The builders of these cells are commonly called mud Wasps. When one of these masons has chosen a place, and has begun to work, she brings in her jaws a lump of soft mud. It is not certain where she got it — whether she gathered some dust and moistened it with the liquid of her mouth, or, as some think, she gathered it where the earth is softened by the wash of the sink. At any rate, she has kneaded it perfectly, and she spreads it as easily as the mason lays his mortar.

Mr. Gosse watched a Dauber, and tells some curious things about her. The first cell was nearly done; the Wasp had just closed the mouth. While gone for more, a pin was thrust through the mud into the cell. When the Wasp came, she laid her mortar over the hole, spreading it very skillfully and evenly. When gone again, the pin made another hole, which she closed up; and so for several times. Finally Madam Wasp got angry, and began to buzz about, trying to catch the house-flies which were near. She seemed quite certain that they had done the mischief, and waited after she had laid more mortar, as if expecting to "catch them at it." Then Mr. Gosse broke off a large piece of the side and bottom, showing the grubs, and the small spiders which she had tucked in for her children's food. This breach

she repaired as quickly as possible, in two or
three loads, laying the mud all round the hole,
and closing up at the middle.

Presently she began to build another cell,
and again she found trouble. A tin-tack was
placed in the mud, just where she would lay
the next load. When she came back, she
seemed quite "bothered;" she ran back and
forth over the cells for some time, with the
mud in her jaws, at a loss what to do. A
hole she could stop up, but here was something in the way. If she should lay the
mortar in its place, the tack would be more
firmly fixed. If she should place it elsewhere, it would be wasted, or might do harm;
if she would try to remove the evil, she must
lay down her burden. At length she seized
the tin-tack in her jaws and pulled it out, drop-
ping the mud as she did it. Next time she
went away, a bit of worsted was pressed into
the mud, which made still more serious trouble, as the bit which she could seize would
yield without coming away. Still, by taking

hold of the different parts, one after another, and tugging at them a long time, and by walking round and round with it in her mouth, she at length pulled it out.

The Dauber Wasp builds the walls of the cell, and lays an egg. Then she finds some spiders of a beautiful green species, and puts them in, bringing them very carefully in her jaws and feet. These she walls up with the egg, and the grub, when hatched, eats up the soft parts of the abdomen.

When autumn comes, the Wasps seek for hiding places in the crevices of houses, where they may pass the torpid months. Sometimes they crawl away where their presence is not desired — into clothing, and between sheets. An acquaintance had a beautiful black pointer dog, named Don. Don had a great dislike for black Wasps, and when they began to creep about, looking for their hiding places, he killed very many of them. He would draw back his lips from his teeth, so that they might not sting him, and then snap them in his teeth, throwing them quickly on the floor. If the Wasp writhed or crawled, another and another snap was sure to follow, until the crushed insect showed no more signs of life.

A large and fierce variety of Wasps is called the Hornet. Its sting is very venomous, and its temper none of the best. It will follow a person, single handed, with great perseverance, when its wrath has been provoked.

Another very tetchy and hot tempered little thing, is a smaller variety, known to school boys as Yellow Wasps. They are usually quiet enough when undisturbed, but woe to the foolish boy who throws a stone, or thrusts a stick into their paper house. The angry swarm issues forth; they buzz about the ruined nest for a moment, and then, discovering the author of the mischief, they fly in solid column to avenge the wrong. If the unlucky urchin has not speedily taken himself far away, he will have good cause to repent an injury to a quiet and unoffending, if not inoffensive, community. These fellows do not give any warning, like the honey bee, but true as an arrow to the mark, they go straight at you, and ear, eye, cheek, lip —the part hit, suffers. The best course for the boy is to pocket the affront, and put some aqua ammonia, also called spirits of hartshorn, on the wound. Better still, let the Wasps alone in the outset. If it is necessary to remove them, put a wisp of straw on the end of a pole, and burn them out at nightfall. If it is desired to remove a nest with the inhabitants, for study, the Wasps may be quieted with chloroform, applied at the bottom of the nest, by a bunch of cotton.

THE MIGRATORY LOCUST.

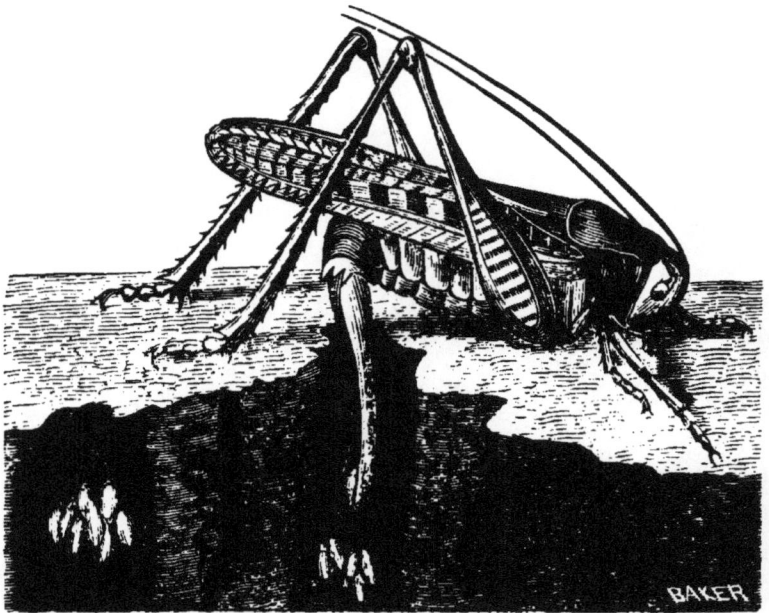

GRASSHOPPER LAYING EGGS.

About Locusts.

ARTICULATA — INSECTA.

ORDER — *Orthoptera.* Straight-winged.

FAMILY — *Locustidæ.* Locust-like.

LOCUSTS and Grasshoppers belong to the same order, and few but naturalists know the differences between them, or are able to distinguish the species of either. They have the same general shape — a long body, stiff, folded, fan-like wings, under straight, hard wing-covers, a head not unlike that of a horse, and long legs, the last pair having long and very strong thighs, with which they leap very far. The Arabs say that the Locust was

made of scraps of all animals. That it has the
head of the horse, the horns of the stag, the
eye of the elephant, the neck of the ox, the
breast of the lion, the body of the scorpion, the
hip of the camel, the legs of the stork, the
wings of the eagle, and the tail of the dragon.
The wings of some are spotted, and the spots
have been supposed to foretell future events.

Locusts have been counted among the most
fearful plagues which have ever punished a
nation. In Eastern lands they have appeared
in astonishing numbers; their swarms have
darkened the sun; they have eaten every green
thing, leaving the land behind them black as if
burned with fire. They are not even content
with that which is green, but devour every
thing which can be devoured — linen, woolen,
silk, leather, the very varnish of the furniture.

In 1748 the locusts appeared early in June
in Hungary, on the Danube. In July they
were terribly destructive throughout Poland,
and at the middle of August they appeared in
clouds in London. In one night they ate the

grass and the foliage of trees about Vienna, making the forests bare as brooms. In Poland they covered the country for miles, and were heaped up a foot thick; when they alighted they covered the ground like falling snow. At Warsaw soldiers were sent out against them with cannon. The firing of great guns scattered them and drove them away. In Italy the government offered rewards for them, and 12,000 sackfuls were gathered, cast into pits, and covered with quicklime.

The prophet Joel gives a description of their coming, both sublime and accurate: " A day of darkness and of gloominess, a day of clouds and of thick darkness, as the morning opened on the mountains; a great people and a strong. A fire devoureth before them, and behind them a flame burneth. The land is as a garden of Eden before them, and behind them a desolate wilderness; yea, and nothing shall escape them. The appearance of them is as the appearance of horses, and as horsemen shall they run. Like the noise of chariots on the tops of mountains so shall they leap; like the noise of a flame of fire that devoureth the stubble, as a strong people set in battle array. They shall run like mighty men; they shall climb the wall like men of war. And they shall march every one his ways, and they shall not break his ranks; neither shall one thrust another. They shall walk every one in his path, and when they fall upon the sword they shall not be wounded. They shall run to and fro in the city. They shall run upon the wall. They shall climb up upon the houses. They shall enter in at the windows like a thief. The earth shall quake before them; the heaven shall tremble; the sun and moon shall be dark, and the stars shall withdraw their shining."

The prophet also mentions the usual way in which the locusts are destroyed: "I will remove far off from you the northern army, and will drive him into a land barren and desolate, with his face toward the east sea, and his hinder part towards the utmost sea, and his stink shall come up because he hath done great things." This prophecy may refer to the coming of an army of human beings, but the description literally applies to the march of these insects as described by historians. Mr. Barron says that in 1784, and in 1797, two thousand miles in South Africa were covered with Locusts, which, being borne into the sea by a northwest wind, formed, for fifty miles along the shore, a bank three or four feet high; and when the wind was in the opposite point, the horrible odor from them was perceptible a hundred and fifty miles away.

Most scourges bring in their train benefits which fully repay, if they do not many fold surpass, the injury inflicted. The prairies rejoice in a greener verdure after the fire has consumed the withered grass. So a land which has been choked with rank shrubs and withered bitter grasses, after it has been swept by the Locusts, soon wears a more beautiful dress, with new herbs, superb lilies, fresh annual grasses, and young and juicy shrubs, which afford sweet pasture for wild cattle and game.

Locusts are eaten by all sorts of quadrupeds, by many birds, large and small, and even by man. In the countries which they ravage, the people have nothing else left to eat, and learning from necessity, they continue to eat Locusts from choice. The Arabs boil them and dry them in the sun. Others soak them in oil. In other places they are gathered in heaps and salted. The wings are taken off, and the bodies eaten as meat, or they are dried, ground, and made into bread. They have even been exported, and armies have been relieved by them. The African Bushman delights in a swarm of Locusts, as his choicest game, furnishing plenty of food without having to work for it. He makes large fires, and the Locusts, flying through the flame, have their wings scorched, fall into the fire, are roasted and eaten. Those that remain are ground between stones, and kept for another meal. Europeans dislike them, but the fault is probably in the cooking; Dr. Livingstone thinks them very good eating when well prepared.

Honey is eaten with them, when it can be had, as it assists digestion.

Does this remind us of John the Baptist, whose "meat was locusts and wild honey." It has been questioned whether the quails which the strong east wind blew together for the Israelites in the desert, were not truly Locusts; there is doubt whether the word translated quail had ever that meaning. The Jews ate Locusts, and distinguished between such as were clean or unclean.

The young Locusts do not pass through the several changes which most insects undergo. The bee, for example, is first a grub, then a chrysalis, then a perfect, winged bee. The Locust comes from the egg a Locust, but wants wings, which come gradually. The eggs are laid in the ground. The female pierces the ground with a long, two bladed, hollow instrument. When it is forced into the soil, the blades open a little, and press the earth aside, while a dozen eggs are passed into the cavity then formed. The contrivance is not unlike a corn-planter, which makes a hole, drops the corn, and covers it, all at once. The Locust goes about thus, planting her eggs, until she has deposited several hundred. They remain during the winter, and until the warm sun next summer hatches them, bringing out little creatures as large as gnats. These stay a while in the nest, and in the ground near by, and then come forth, hopping about without wings. As they grow they shed their skins, each time appearing in a new, larger, and more perfect dress. By the third or fourth change, wings begin to appear, and by the sixth they are full fledged. The common Grasshoppers make their entire growth in one season, but the terrible migratory Locust, which has been mentioned above, is said to live in the ground two years, and come forth in the third.

It is often a matter of surprise that insects like the Locusts, the chinch-bugs, and others, should not be observed for many years, and then should appear in swarms of such immense numbers, and do such terrible mischief. Many

attempt to account for this by supposing that
the ground has some hidden power of sponta-
neous production, which is thus fitfully exerted.
It is probably the fact that these insects never
entirely disappear; that no season passes without producing enough to keep up the succession. They are exceedingly productive, so that
a few may be the parents of a multitude. But
the dangers which surround the eggs and the
young, eaten as they are by every kind of bird
and insect, and destroyed by myriads by unsea-
sonable cold and rain, sweep them away, and
leave only a remnant for seed. If only one in
a thousand escapes, that one will reproduce a
thousand. Thus if two favorable seasons follow in succession, the scourge appears, and the
crops suffer.

In the south of Europe rewards are regu-
larly paid for the collection of Locusts and of
Locusts' eggs. The city of Marseilles expended
20,000 francs for that purpose, in one year. A
franc is paid for about two pounds and a quarter of eggs. In Italy large quantities have been

gathered and thrown into the streams. There is a slight difference in the piercer of the Locust and of the Grasshopper, but the method of placing the eggs in the ground is essentially the same.

One of the most noted among the Grasshoppers is the Katy-did. This insect is of a pale green color; its head seems to have been squarely chopped off; its wing-covers are rounded, and enclose the wings and body like the sides of a pea-pod. It lives in the branches of trees, and does not lay its eggs in the ground, but deposits them upon the twigs and branches in regular rows. The song of the Katy-did is one of the cheerful sounds of autumn, save that from constant repetition it becomes tiresome. It is not truly a song, for it is not made by the mouth.

" The musical organs of the male consist of a pair of taborets. They are formed by a thin, transparent membrane stretched in a strong, half-oval frame in the triangular overlapping part of .each wing cover. During the day they are silent, but at night the males begin the joyous call by which they enliven their silent mates. This proceeds from the rubbing of the taboret frames against each other when the wing covers are opened and shut; and the notes are repeated for hours together. The sound may be heard in the stillness of the night to the distance of a quarter of a mile. At the approach of twilight the Katy-did mounts to the upper branches of the tree in which he lives, and, as soon as the evening shades prevail, begins his noisy babble, while rival notes issue from the neighboring trees, and the groves resound with the calls 'Katy-did-she-did, she-didn't, she-did,' the livelong night."

He put his acorn helmet on;
It was plumed of the silk of the thistle-down;
The corslet plate that guarded his breast,
Was once the wild bee's golden vest;
His cloak, of a thousand mingled dyes,
Was formed of the wings of butterflies;
His shield was the shell of a lady-bug queen,
Studs of gold on a ground of green;
And the quivering lance which he brandished bright,
Was the sting of a wasp he had slain in fight.
Swift he bestrode his fire-fly steed;
  He bared his blade of the bent-grass blue;
He drove his spurs of the cockle seed,
  And away like a glance of thought he flew.

The moth-fly, as he shot in air,
Crept under the leaf, and hid her there;
The katy-did forgot its lay;
The prowling gnat fled fast away;
The fell mosquito checked his drone,
And folded his wings till the fay was gone;
And the wily beetle dropped his head,
And fell on the ground as if he were dead.

They watched till they saw him mount the roof
  That canopies the world around;
Then glad they left their covert lair,
And freaked about in the midnight air.

                 The Culprit Fay.

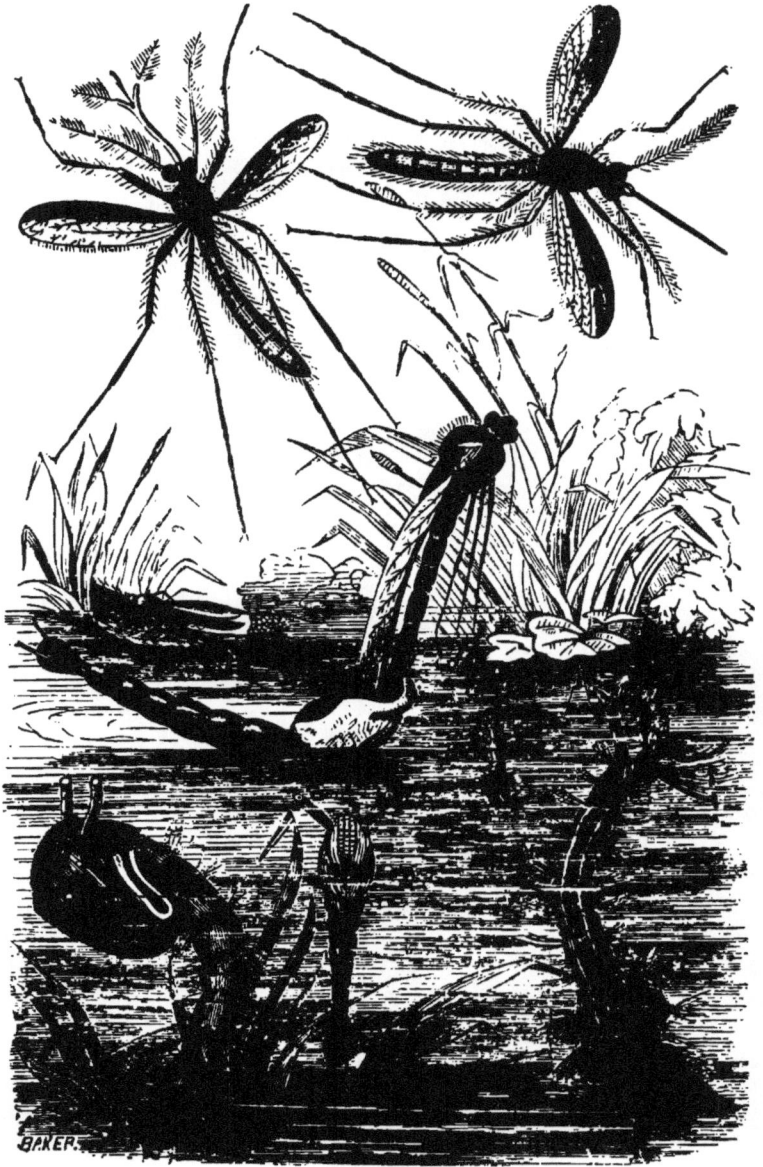

TRANSFORMATIONS OF THE MOSQUITO.

About Mosquitoes.

ARTICULATA — INSECTA.

ORDER — *Diptera.* Two-winged.

FAMILY — *Culicidæ. Culex,* A gnat.

THE Mosquito is a nuisance. He sings and then he bites; and his singing is usually notice that he intends to bite. He comes in the night, when the faint and sultry air persuades the sleeper to throw off the protecting cover, and sleep flies before him. If one seeks a shelter from the glaring sun, under the shade by the brook-side, myriads of these gauze-winged musicians warn him away from their realm. The wild forest is full of them. The heavy timber, from June to September, is utterly uninhabitable, unless constant war is made against the Mosquitoes.   Every bit of standing water, and every purling rill, teems with them. The trees and bushes every where shelter them. Shake a bough, and a swarm rises from it; land from a boat, and a cloud tender a too cordial reception.   They gather like hungry politicians about the dispenser of official favors.

Nothing but thick leather and woolen will protect your ancles or your wrists.   You can save your face in only one way.   Wear a soft hat that you can sleep in; get a yard of black lace, sew the ends together, and draw round the crown of your hat one end of the bag which you make, while you gather the other end under your chin. The brim of the hat will keep the veil from the face, and the disappointed Mosquitoes will rave in vain against the outside.   In a few moments one becomes so accustomed to the veil that it does not interfere with sight, although it is in the way in eating and drinking.   It is no exaggeration to say, that in the dense forest, in June, the Mosquitoes have gathered upon the back of a man sitting down to rest, so thickly as to hide the color of his coat, whether light or dark, with the brown of their wings.

The Rev. Walter Colton tells how the miners in California made culprits disclose the truth by means of Mosquitoes. A rogue had stolen a bag of gold and hid it. Neither persuasions or threats could make him tell where it was concealed. He was sentenced to receive a hundred lashes, but was told that he would be let off with thirty, if he would tell what he had done with the gold. He refused. The thirty lashes were laid on, but he was as stubborn as a mule. He was then stripped, and tied to a tree. The Mosquitoes, with their sharp bills, went at him, and in less than three hours he was covered with blood. Writhing and trembling from head to foot, he exclaimed, "Untie me, untie me, and I will tell where it is!" "Tell first," was the reply. So he told where it might be found. Then some of the party, with wisps, kept off the hungry Mosquitoes, while others went where the culprit directed, and found the bag of gold. He was then untied, washed with cold water, and helped to his clothes, while he muttered, as if talking to himself, " I couldn't stand that, anyhow."

There is no doubt that a man would perish in a short time, from loss of blood, and from the fever caused by the poison of their bites, if exposed, as this man was, with no means of defence.

The largest kind of Mosquito about the Mississippi river is called the " Gallinipper." The boatmen say that it is as large as a goose, and that it carries a brickbat under its wing, on which to sharpen its bill.

Cattle are not troubled by Mosquitoes, but horses suffer terribly. The lumbermen drive them away from their camps by making low fires of chips and damp grass. In the coolness of evening, the smoke from these " smudge " fires hangs heavily over the ground, and affords considerable protection, which even animals seek.

On a still night such a camp is very pictur-
esque. The low log hut by the river; the tall,
sombre pines, towering above dense masses of
maples, and ragged outlines of oaks; the strag-
gling fires, that thrust out tongues of fitful
flame, and reek with thick smoke, which
spreads upon the ground, or lazily rolls over
the roof; the long, level lines of blue haze
which the smoke finally draws against the foliage of the trees; the solemn stillness resting
over all, broken only by the hoot of the owl,
the wail of the whip-poor-will, or the tinkle of
the rippling stream, while the bright eyed
climbing stars replace the waning twilight;
compose a scene too lovely to be spoiled by
millions of myriads of swarming, howling, rav-
ing, hungry Mosquitoes.

The Mosquito is an insect of the water.
Early on a summer morning, even before sun-
rise, the mother may be found laying her eggs.
They must be placed where there is warmth
enough to hatch them, and where the young
creatures which pop out may go at once into

the water.   So the careful insect, like the
mother of Moses, puts her children into a little
ark, which she leaves on the surface of the pool.
The ark she makes of the eggs themselves.
She rests on a bit of grass, or a leaf, at the top
of the water, holding to it by the first and second pair of legs.   The third pair she crosses
behind her like the letter X.   The first egg is
caught and held between the legs.   Then
another and another are fastened to the first
by the gum which covers them, until fifteen or
twenty have been set up side by side, as one
might set up a number of ears of corn, or like
the seeds in the head of a sunflower.

When the mass becomes too heavy for her
to support, she lowers it upon the water, but
still holds it by putting her feet on either side,
until two hundred and fifty or three hundred
eggs have been laid.   Those at the sides are
higher than those in the middle, while those at
the ends are raised somewhat more.   Thus the
whole mass is shaped much like a canoe.
These tiny black boats, about as large as grains

of wheat, may be found floating upon the top of any tub, or barrel of water, which has stood for some days. Nothing can harm them, if some other creature does not eat them up. The storm may dash them against the shore, but they are too light to break ; a torrent of water may be poured upon them, and they come out of the bubbling foam as buoyant as air, and as dry as a duck; the water may freeze solid, but their life is not destroyed.

In a few days — three are usually enough, if warm — the eggs hatch, and each sends a wriggler down into the water, through a hole in the bottom. The little fellow swims about, and presently hangs himself by his tail to the surface. If disturbed, he goes down out of the way, but soon comes back, and rests, as before, with the tip of his tail out of water. He does this, just as other swimmers do, because he would keep his nose above water. The odd thing about it is, that his nose, or, at least, the tube which he breathes through, is not on his face, but at the tip of his tail. It ends in a few hairs, which spread in a star-form, and are oiled, to repel the water. Thus the tail is both nose to breathe through, and buoy to keep itself at the top of the water. He lives upon the impurities in the water, and so serves a very useful purpose in the world.

By and by, he changes into a pupa, and then he turns himself over, end for end. He did breathe through his tail; now he breathes through his ears, or a pair of tubes which look like ears, and are thrust up, just a little, out of the water. His tail is now like the tail of a fish, and by it he can move himself through the water as he pleases. He remains thus about fifteen days, and then takes a new form, exchanging his home in the water for a life in the air.

When the warm sun shines on the water, the change comes. The pupa rises to the surface, and thrusts out his head and shoulders. The cover bursts, and the plumed head appears, followed by the shoulders, and the filmy wings. Now is the time of danger. If an unlucky puff

of air sweep the water, over goes our sailor, his
wings are wet, and his voyage lost, just as he is
ready to come into port. ·His old garment lies
upon the water. It is his life-boat. His body
is the mast, and his drying wings are the sails.
Now his slender legs are dry, and with them he
feels for the surface of the pool. He lifts himself free from his cast-off coat, rests an instant
on the water, and then leaps into the air, a singing, stinging Mosquito.

But all the Mosquitoes do not sting. The
males wear a pair of plumes upon their heads,
and spend their days in a ceaseless dance in
the sunbeams. Those that bite are the females.
One gently drops on your neck or hand, with
footstep so light that you feel it not; she looks
about for a moment, hesitating as to where she
will begin to bore. Now she has found the
place, and her needle tongue goes down into
the skin. Now you feel the prick, and now
you may see her chest heave as she pumps up
the red fluid. No speculator, boring for oil,
ever felt happier over a flowing well, than our

borer over the flowing fountain which she has tapped. Now her abdomen expands, more and more, until it seems that she will burst. At last she has enough — too much, in fact, for her greed will cost her life. She draws up the rod, and heavily flies away. Her light wings can scarcely bear the increased burden. She will die of surfeit.

TO A MOSQUITO.

Fair insect! that, with thread-like legs spread out,
  And blood-extracting bill and filmy wing,
Dost murmur, as thou slowly sail'st about,
  In pitiless ears full many a plaintive thing,
And tell how little our large veins should bleed,
Would we but yield them to thy bitter need.

Unwillingly, I own, and, what is worse,
  Full angrily men hearken to thy plaint;
Thou gettest many a brush and many a curse,
  For saying thou art gaunt, and starved, and faint:
Even the old beggar, while he asks for food,
Would kill thee, helpless stranger, if he could.

Beneath the rushes was thy cradle swung,
  And when, at length, thy gauzy wings grew strong,
Abroad to gentle airs their folds were flung,
  Rose in the sky and bore thee soft along;
The south wind breathed to waft thee on thy way,
And danced and shone beneath the billowy bay.

Calm rose the city spires, and thence
  Came the deep murmur of its throng of men,
And as its grateful odors met thy sense,
  They seemed the perfume of thy native fen.
Fair lay its crowded streets, and at the sight
  Thy tiny song grew shriller with delight

At length thy pinions fluttered in Broadway—
  Ah, there were fairy steps, and white necks kissed
By wanton airs, and eyes whose killing ray   .
  Shone through the snowy veils, like stars thro' mist;
And fresh as morn, on many a cheek and chin,
Bloom'd the bright blood thro' the transparent skin.

Sure these were sights to touch an anchorite!
  What! do I hear thy slender voice complain?
Thou wailest, when I talk of Beauty's light,
  As if it brought the memory of pain.
Thou art a wayward being — well — come near,
And pour thy tale of sorrow in mine ear.

What says't thou—slanderer! rouge makes thee sick?
  And China bloom, at best, is sorry food?
And Rowland's Kalydor, if laid on thick,
  Poisons the thirsty wretch who bores for blood?
Go! 'twas a just reward that met thy crime—
But shun the sacrilege another time.

That bloom was made to look at, not to touch;
  To worship, not approach, that radiant white;
And well might sudden vengeance light on such
  As dared, like thee, most impiously, to bite.
Thou should'st have gazed at distance, and admired,
Murmured thy adoration, and retired.

                                        Bryant.

SCARABÆI AT WORK.

ABOUT BEETLES.

ARTICULATA.— INSECTA.

ORDER — *Coleoptera.* Sheath-winged.

KNIGHTLY armor of proof protects the Beetle. First, there is the strong helmet, with shut visor, and crest of varied device. Then comes the solid cuirass, which protects the body, and below that the full-orbed, or oval shield, which covers the abdomen, and the upper joints of the legs. He carries neither sword nor lance, mace nor battle axe, but from the joints of his visor project two ponderous jaws, which grip like a vice. He is horse and horseman in one. His thick shield parts in the middle, and when the two leaves swing apart, they disclose a pair of light, gauzy wings, which, with a great deal of fussy buzzing, lift him from the ground and carry him away, when he,

"Drowsy beetle, wheels his droning flight."

The plates of his coat of mail fit each other very exactly. The helmet makes the neatest joint with the corselet, and the corselet with the shield. The wearer can move every part with perfect freedom, and yet each joint is closed against prick of arrow or thrust of spear. Yet the Beetle is not the swift horseman of to-day, but resembles more the heavy man-at-arms of three hundred years ago. When he was pushed off his horse in sham or real fight, and lay sprawling on his back, boxed up in his heavy plate armor, he needed a stout esquire to set him on his pins again. Just so, if a bumming beetle be knocked on the floor; it takes him a long while to overcome his astonishment, and make ready again for a tilt at the lamp, or at your face.

Our knight has little of the swift dash of the wasp, who pricks with his sharp lance, and then rings his shrill defiance. He has none of the stealthy adroitness of the spider, who lassos his victim, like a Mexican, and then stabs him in the back, as coolly as an Italian bravo. Indeed, he does very little at offensive warfare. If you are in his way, he gives you a sharp pinch, or whacks you in the face, but that is all. His heavy mail serves mostly to ward off the assaults of others.

The style and the ornaments of his armor are very various, and often are very beautiful. Sometimes the whole suit is plain black, or dark brown. Sometimes it gleams with brilliant hues of green, crimson, purple, and gold, or blazes with precious gems, set in polished metal. In any case, he keeps his armor scrupulously clean, no matter how filthy the work which he is busied about.

The order contains over one hundred thou-

sand kinds, divided into various families. We
must be content with noticing a few of the most
remarkable. Some of them do great injury to
vegetation, either while grubs, as the borers in
trees, or the young of the cock-chaffer, which
eat the roots of grass; or while fully developed
beetles, as the curculio, which kills the plums,
the striped cucumber-bug, the rose-chaffer, and
many others. Other kinds confer decided benefits. The Water Rovers, the Skin Beetles,
Carrion Beetles, and Dung Beetles, are scaven-
gers, disposing of the filth in which they and
their young live. The Tiger Beetles, Lady-
birds, and Diving Beetles, prey upon caterpillars, and plant lice. The Stag Beetles, Bark
Beetles, and others, help destroy old trees
which are going to decay. The Blister Beetles,
or Cantharides, are pounded up by the druggists, and the dust is spread upon plasters, to
raise blisters when applied to the skin.

The first on our list is the Sexton, or Burying
Beetle. If the body of a dead bird, or mouse,
or any piece of meat, be left upon a spot of soft

earth, it will often be found, on the next morning, half sunk in the soil. Take up the bird, and you will find under it one or two beetles, sometimes entirely black, sometimes barred with orange. During the day the insects will usually be quiet, but at nightfall they will begin work again. The work of burying is done almost entirely by the male Beetle, the female either hiding in the dead body, or sitting quietly on it, and being buried with it. The male begins by turning a furrow all round the bird, about half an inch away. His head is held sloping outwards, and like a strong plow, turns the earth aside. When the first furrow is made, a second is turned within it, the dirt being thrown into the first. Then a third is made, and this is quite under the bird, so that the Beetle is out of sight. The work may be traced by the heaving of the earth, which now makes a wall, and as it grows higher, the bird sinks. After hard work for about three hours, the Beetle comes forth, and crawls upon the body, to see how he succeeds. He rests half an hour, goes down again, dives into the grave, and pulls the bird down by the feathers. He works two or three hours more, plowing and pulling ; then comes up, takes another survey, and drops down, as if suddenly fallen asleep. When he is rested, he pulls the bird about, this way and that, tramples it down, and settles it to his mind. Then he goes behind the rampart of earth, and plows it back into the grave, with great skill and strength. He bends his head down first, and then turns up his nose with a jerk which throws the earth forward. When the grave is filled, and carefully examined, no feather being left in sight, he digs a hole in the loose earth, and having already buried the bird and the female, next buries himself. The female lays her eggs, the pair take a full meal of the carcass, then dig their way out, and fly away.

If the creature is no bigger than a mouse, a single day will be long enough to bury it in. One buried a mole, forty times as large as itself, in two days. A French naturalist placed two pairs of these Beetles under a glass case, and furnished them with dead bodies. In fifty days they had buried four frogs, three small birds, two fishes, one mole, two grasshoppers, and three bits of flesh. All this work is done to secure a nest and food for the young, and to protect it from other creatures, as the fox, or the raven, which might devour flesh and young Beetles together.

The work is done very much as men sink wells in sandy ground. They sometimes lay down on the earth a ring of plank, as large as the well is to be. Then they build a circular wall of brick and mortar upon the ring. The sand is taken out from under the plank, and the whole wall sinks slowly down. So the well is dug as deep as may be necessary, while the wall is built up at the top, as fast as it settles into the ground.

Another burying Beetle is the Dor Beetle, called, in this country, Tumble-bug. This is akin to the sacred beetle of ancient Egypt, or the Scarabæus. Its image was engraved on rings, which soldiers wore to show that they were warriors. On temples or columns it was a symbol of Divine wisdom, which regulates the universe, teaches mankind, and is self-existent. In its singular habit of rolling about pellets of dung the Egyptian astrologers thought it represented the revolutions of the sun, moon, and stars.

When this insect finds a patch of cow-dung, she sets herself at work. First she digs a deep hole, smooth and round. Then she cuts off a portion of the patch, lays an egg in it, and rolls it about into a rude ball. Now she rests her fore feet upon the ground, and with her hind feet rolls the ball hither and thither, until the outside has gathered in the dust and sand, a thin, hard crust or shell. Then, always pushing backwards, she rolls it to the hole, tumbles it in, and covers it with earth. The egg is soon hatched, the grub feeds on the substance which surrounds it, changes to a chrysalis, and remains in the shell, which still serves to protect it, until it is ready to come forth, a perfect Beetle,

qualified to roll pills on its own account. The
smooth surface of this beetle, which gleams like
polished steel, retains no trace of its work.
Not a spot or a stain defiles it, nor does any
odor betray its occupation.

Some of these rolled cocoons are very large.
A specimen in the British Museum, made by
the *Goliath*, is as large as a swan's egg. The
walls are quite thin, and are strengthened by a
belt about the middle. The insect which grew
in it is still inside, and may be seen through an
opening at one end. Its plates of mail are rich,
velvety chocolate, edged with broad bands of
white. This species, the largest now known,
has a body about four inches long and two
inches broad; when walking, it covers a space
of nearly six inches.

An interesting family of Beetles includes the
Dors, May-bugs, Cock-chafers, and Rose-bugs.
They are very common and well known, both
as Beetles and as grubs. The perfect insect
does not live more than a week, and the species
is not seen more than four or six weeks in a

season. The females burrow in the ground about six inches, to lay their eggs — some say, as many as two hundred. In about fourteen days the eggs are hatched, and produce white grubs, each having six legs near the head, and a pair of strong jaws. Their skins are white, and partly transparent. When thrown out by the plow or spade, they are found coiled like a ring, or a horse-shoe. A full grown grub is nearly as large as one's little finger — a plump, fat morsel, very eagerly swallowed by crows and chickens.

They do much mischief, eating the tender roots of grass, grain, herbs, and trees. When very numerous, they have so cut off the roots of grass, that the turf could be rolled up in many places like a carpet. As winter comes, they descend below the reach of frost, and lie torpid until spring; then they change their skins, and go back to the surface. At the end of the third summer — some say the fifth — they go down about two feet, and each one, by moving from side to side, forms a hollow oval space, about the size of a pigeon's egg. Here it casts its skin, and becomes a pupa, whose clear salmon-colored skin shows under it the head, eyes, jaws, wings, legs — all the parts of the perfect beetle. In February this skin bursts, and the insect is ready to dig its way out when the first warm week of May has clothed the trees with leaves. In digging out decayed stumps of trees, fine opportunities may be had for observing these grubs in every stage of their lives.

The winged beetles do as much harm by eating the leaves of trees, as the grubs did by devouring the roots. During the day they remain on the branches, hidden under the leaves. At nightfall they begin to buzz about, humming among the trees until midnight. They often come into houses, attracted, and blinded, by the light. They dart about with very uncertain aim, putting out candles, whacking unlucky people in the face, and banging against trees and walls so hard as to throw themselves stunned upon the ground. Hence come the sayings "blind as a beetle," and "beetle-headed." When very plenty, the attempt has been made to check their ravages by shaking them from the trees upon cloths spread underneath. They are then thrown into boiling water, and fed to fowls or hogs. In this way they have been gathered by pailfuls, and in a few days no more could be found. Many years ago these beetles were so plenty in Norwich, England, that a farmer and his men claimed to have gathered eighty bushels of them; and they and their grubs had done so much harm, that the city gave the farmer twenty-five pounds for relief.

The Rose-chafer, like the May-bug, does much harm in gardens and nurseries. It is about one third of an inch long, covered with yellowish down. The slender, red legs end in long feet. They come forth about the second week in June, and remain about a month. The eggs are hatched in the ground, and the grubs feed upon roots until autumn, when they descend below the frost. In the spring they

come up again. In May they pass the first
change, and in June come forth fledged Rose-
bugs. They can be destroyed only by crush-
ing, burning, or throwing into scalding water.
They eat the leaves not only of rose bushes, but
also of fruit trees.

Certain Beetles called Stag Beetles, and
Horn-bugs, from their jaws, which resemble
the horns of oxen or deer, belong to the family
Lucanidæ. They fly by night, and often the
lights attract them into houses, to the great
alarm of the people within; but they are quite
harmless, and will not even pinch, unless provoked. Their grubs resemble those of the
Scarabs, and live in the trunks and roots of
trees. When full grown, they make cocoons
of bits of wood and bark, glued together, and
wait the changes which make Horn-bugs of
them.

The grubs of the *Buprestidæ* are the borers
so destructive to fruit and forest trees. They
are long, narrow, and flat, with large, hard
heads and jaws. They have no legs, but

move by twisting from side to side, and by pulling with their jaws. They may be destroyed by thrusting a wire after them, into the hole which they are making — if it can be found. The crushed grub will come out on the point of the wire, to prove the success of the experiment. Possibly, however, woodpeckers would be more expert at this kind of thing than men.

Another family that are great pests are the *Curculios.* These spoil the fruit, attacking plums, apricots, and cherries, and not sparing apples, pears, and peaches. As soon as the fruit is set in the spring, these little insects begin their work, and they keep at work, until July or August. The Beetle cuts with its snout a little curve in the skin of the plum, then turns round and lays an egg in the wound. A maggot hatches, which eats its way into the fruit, even to the stone; this causes the plum to become diseased, and to fall off before it is ripe. When the plum is partly grown, the curved scar may be easily found. All such, with all that fall upon the ground, should be gathered and destroyed, to prevent the maggots from going into the ground to pass the changes, and coming out afterwards to keep up the evil.

Others of this family attack the pines. Wilson says : " Would it be believed that the larvæ of a fly no bigger than a grain of rice, should silently, and in one season, destroy some thousands of acres of pine trees, many of them two and three feet in diameter, and a hundred and fifty feet high! In some places the whole woods, as far as you can see around you, are dead, stripped of their bark, their wintry arms and bare trunks bleaching in the sun, and tumbling in ruins before every blast." Besides boring into the trunks, these insects often destroy the top shoot of the tree, on which its straight and lofty growth depends. Mr. Wilson suggests that until farmers can devise some better plan of killing these pests, they had better thank and protect the woodpecker.

Another rascal of this tribe is the wheat-weevil. Several insects are known by this name, but the one meant is a slender Beetle, about one eighth of an inch long, and of a pitchy-red color. It lays eggs upon the harvested wheat, and the grubs burrow into the grain, each one taking possession of a single kernel. The worm eats the substance, but leaves the shell, so that the only evidence of harm is the lightness of the grain. They may be destroyed by drying the wheat in kilns.

After the peas have blossomed, and while they are just beginning to swell in the tender pods, the Pea-beetles gather by night, or in cloudy weather, and lay their tiny eggs in minute holes which they pierce in the surface of the pods. The maggots, as soon as hatched, bury themselves in the peas, and the small holes are soon closed. Often every pea in a pod contains a grub. They remain in the peas after they ripen, and come out in the spring, perfect bugs, ready to carry on the work. Those who plant "buggy" peas will find the bugs are quite as sure to come up, and bear fruit, as the peas. The crow-blackbird is fond of the bugs, and the Baltimore oriole, or hang-bird, splits open the pods to get the grubs in the green peas. Don't disturb him, for be sure if he wants the peas, you don't.

Another nuisance in the garden is the yellow striped bug, which destroys the young cucumber and melon plants. It comes as soon as the young vines come, and it stays all summer. Great numbers visit the flowers of the squashes and pumpkins for the pollen. The eggs are laid, and the grubs grow, in the ground. Those bugs which have their heads pinched off will be sure to do no further harm. Various devices have been employed against them, such as sifting soot, snuff, sulphur, ashes, or plaster of Paris, on the vines; sprinkling them with steeping of tobacco, red pepper, walnut leaves, or hops; burning fires of pine knots or bits of tar barrels at night, but none are sure. The best preservative is a frame of board, covered with millinet, to place over each hill of vines. If the plants can be protected for a little time, they will grow fast enough to escape further harm.

But my readers will begin to believe that all the Beetles are nuisances, and will be led to make war upon every insect which wears a hard shell. Let us find some for which something may be said on the credit side of the account.

While writing about ants, we mentioned those destructive little insects, which the ants use like herds of cattle, the plant-lice, or *Aphides*. The ants only milk their cows, but the grubs of a Beetle eat them up. These grubs become the pretty insects which people commonly call Lady-birds; naturalists call them *Coccinellidæ*. They have the size and shape of half a pea. Some are black, spotted with red; others, red with black, yellow with black, or yellow with white spots. The eggs are laid among the lice, and the grubs at once go to work catching and eating prey as large as themselves, without ever seeming to be satisfied.

This insect has always been a favorite. The

German children think it brings fair weather, and the English boys and girls are afraid to hurt it, lest it should bring rain. The Norwegians call it Marspäert, and count the spots to see if there will be a good harvest. If there be fewer than seven spots, they say bread will be plenty and cheap. The children sing:

> Marspäert, fleeg in Himmel!
> Bring my'n Sack voll Kringeln; my een, dy een,
> Alle lütten Engeln een.

> Marspäert, fly to heaven!
> Bring me a sackful of biscuits; one for me, one for thee,
> For all little angels one.

The Scotch children call this insect **Lady Lanners**, or Landers. They say:

> Lady, Lady Lanners,
> Lady, Lady Lanners,
> Tak' up your clowk about your **head,**
> And flee away to Flanners.
> Flee ower firth, and flee ower fell,
> Flee ower pule and rinnan well,
> Flee ower muir, and flee o'wer mead,

Flee ower livan, flee ower dead,
Flee ower corn, and flee ower lea,
Flee ower river, flee ower sea,
Flee ye east, or flee ye west,
Flee till him that lo'es me best.

There are many other little rhymes in various languages, which show how children every where love this insect. Perhaps they do not know that it is useful as well as pretty.

The family of *Cicindelidæ* are called Tiger Beetles, or Sparklers. They get the first name from the fierce way in which they seize and devour other insects, and the last from their brilliant colors. The Tiger Beetle is among insects what the kite is among birds, or the shark among fishes. He runs with great speed; he is armed with jaws like sickles, crossing each other; his eyes project from each side of his head, that he may see every way; his wings help him to fly as swiftly as a wasp. His suit of mail, of burnished steel embossed with gold, is more beautiful than any thing ever wrought by mortal armorer. If placed under the microscope in a strong light, his whole surface seems ablaze with precious metals and dazzling gems. The larvæ make their homes in the ground in tunnels about a foot deep. Here one of them lies in ambush just at the top of the ground, the head hooked to the edge of the hole. When an insect passes, the jaws grasp it, and drag it to the bottom of the den, to be eaten.

Another family of carnivorous Beetles lives in the water. Their breathing tubes open under the wing-cases. When they dive, they carry down under the wings a supply of air, and as this becomes exhausted, they rise, lift the wings above the surface, and so take a fresh supply. The larva of the Water Beetle is as active and as fierce as that of the Tiger Beetle, and the full grown insect does not outgrow his youthful tastes. If several be put in a vessel together, they will surely eat each other. A gentleman placed a pair in his aquarium in order to observe their habits. He succeeded in observing, on the next morning, that the male had been killed and partly eaten by his disconsolate widow.

The whirligigs, that shoot from side to side on the top of still water, belong to this family.

One of the most noted Beetles is the Cucuyo, or Fire-fly of, Mexico and Brazil. It wears on each side of the chest two light patches, which by day are pale yellow, but by night glow with a very intense light. When it spreads its wings, its whole body seems filled with the most brilliant flame. It flies by night, and the forests, filled with these insects, crossing and recrossing in every direction, glowing and vanishing as if suddenly lighted and as suddenly extinguished, present a scene too beautiful to be described.

The Indians catch these beetles by balancing hot coals in the air at the end of a stick, to attract them, which proves that the light which their bodies diffuse is to attract. Once in the hands of the women, the Cucuyos are shut up in little cages of very fine wire, and fed on fragments of sugar-cane. When the Mexican ladies wish to adorn themselves with these living diamonds, they place them in little bags of light tulle, which they arrange with taste on their skirts. There is another way of mounting the Cucuyos. They pass a pin, without hurting them, under the thorax, and stick this pin in their hair. The refinement of elegance consists in combining with the Cucuyos, humming-birds and real diamonds, which produce a dazzling head-dress. Sometimes, imprisoning these animated flames in gauze, the graceful Mexican women twist them into ardent necklaces, or else roll them round their waists, like a fiery girdle. They go to the ball under a diadem of living topazes, of animated emeralds, and this diadem blazes or pales according as the insect is fresh or fatigued. When they return home, after the *soirée*, they make them take a bath, which refreshes them, and put them back again into the cage, which sheds during the whole night a soft light in the chamber. In the full glow of one of these Fire-flies, it is easy to read a letter or a book. The little Flies which dart through our meadows in moist summer evenings, are akin, though far less brilliant.

The last family we will mention, are the *Cantharides*, or Blister-flies. They secrete a substance which, when procured by itself, looks like fine snow-flakes; when it is left upon the skin it causes great irritation, and soon produces blisters. The Spanish Fly is nearly an inch long; its color is a satin green, glossed with gold. It feeds upon the ash and lilac, and is found also on the poplar, the rose, and the honeysuckle. Large quantities are taken, killed by fumes of vinegar, and exported for druggists' use. Several kinds of Blister-flies live among us. The Potato-fly, which consumes the vines at midsummer, is of this family. Another often strips the leaves from the clematis. These flies may be caught by shaking them from the vines into water, which prevents their flying, and when dry they may be used by the apothecaries.

AMPHRISIUS BUTTERFLY, CATERPILLAR,
AND CHRYSALIS.

ABOUT BUTTERFLIES.

ARTICULATA — INSECTA.

ORDER — *Lepidoptera.* Scale-winged.

"UGH! See that horrid, ugly worm!" Who has not heard such an outcry? Is there any good reason for the feeling which it indicates? We believe that the repugnance which very many really feel towards creatures of this kind is not, as they think, natural, or inborn, but is the result of early training. When the young mother sees her toddling baby busily watching a caterpillar, she bids him, with earnest words, with looks and accents of disgust, avoid the " horrid, nasty

thing;" his growing curiosity is checked, and
darling Willie Winkie comes to believe that a
worm or a spider is the vilest thing he can
know, as confidently as he believes he loves his
mother or his sister. Whoever has overcome
the feeling, thus artificially acquired, long
enough to begin the study of the forms, the
nature, and the wonderful transformations of
caterpillars of every kind, has learned that in
this, as in all other departments of nature, the
infinite resources of the creative power of God
are wonderfully displayed.

Considering the entire round of the creature's
life, the whole world of birds, insects, and flowers presents nothing more interesting or lovely.
If nature's course is not disturbed, the worm
will fly on wings of beautiful form, exquisite
coloring, and most delicate plumage; the moth
or the butterfly assuredly was, at some day not
long since, a crawling worm. But we go yet
farther, and confidently assert, that at no stage
of its varied life does the insect show to the
student so much that has interest or value, or

to the general observer much more of absolute beauty of color, symmetry, and adaptation, than when it is so often abhorred as a "horrid, ugly worm."

We do not deny that caterpillars of all kinds do much mischief. They eat, eat voraciously, and have the instinct to select the choicest parts of that on which they thrive. Most subsist on vegetable food, and chiefly on leaves; yet some devour the solid wood, some live in the pith, and some eat only grains and seeds. Some kinds attack woolens and furs; even leather, meat, wax, flour, and lard, nourish particular kinds of caterpillars. There is, then, no reason why they may not be destroyed, so that their numbers may be kept within reasonable limits. But we should not assert that the poor creatures are ugly, and then kill them because we have given them a bad name.

Let us see what we can learn by studying the lives of a few; we could wish that every reader, young or old, could have the specimens under his own eyes, sure as we are that he would find more of interest in them than we describe.

In the month of June, when the feathery carrot leaves are growing well, we may find feeding on them a small worm, nearly black, which has perhaps grown to half an inch before we discover him; he may be no more than a tenth of an inch long. If none are seen on the carrots, we may search the parsnips, the leaves of the celery, parsley, or carraway, for the worm thrives on either. He is about as large as an oat-straw, and a little thickest just behind his head. He wears a clean, tight-fitting coat of black velvet, with a broad white band across the middle of his back, and another over his tail; the velvet seems to be laid over him in folds, and to be studded with small black points. If touched, he throws his head back quickly, as if annoyed at the impertinence. Tickle him with a straw, and he pushes an orange-yellow horn out from the top of his head, toward the side which was touched; tickle the other side, another appears. Both issue from the same opening, and the two branch like the two parts of a V. They are scent organs. Immediately a smell is diffused, at first not unfragrant — like some kind of over ripe fruit — but soon sickening; by this odor he probably protects himself from the ichneumon-flies, which would else trouble him; and by it, also, you may know that your specimen is that which we describe.

You may gather a few leaves of the carrot, with the worm, and put them in any safe, airy place where you can watch him day by day; a supply of fresh food will keep him from going away for the present; or you may observe him on the plant where you found him.

In a few days he will quite likely cease to eat. If it were a canary, or a squirrel, which does not dispose of his rations, you might guess that your pet is sick, and so be anxious about him, but you need take little thought for the worm. He becomes restless. He twists quickly from side to side. Presently his skin bursts just above or behind his head, and he actually begins to creep out of it. There, it is done. Your worm is yonder, in a new velvet jacket, several sizes larger, quite differently and more handsomely marked. It is arranged in cross-way folds, as before. On each fold the sober black is enlivened by several bright orange spots; on the middle of the back, where the white fold lay, is a small white spot, surrounded by six others, while three more are arranged a little lower on either side. The old garment, a shriveled, useless thing, lies there, where he crept out of it, after having fastened its hinder hooks to the leaf on which he rested.

Now he takes his food with renewed relish. He moves more freely, and seems much more at ease in his new and enlarged garment. For several weeks this process goes on. He eats, grows, outgrows his old clothes, and creeps out of them in a new and larger suit,— mamma, did you never wish Bobby could do so too, instead of wearing his trousers out at the knees, and kicking his toes through the copper?—

until after four or five weeks, and about as many changes, he is a full grown worm, or caterpillar. When at rest, he is rather more than an inch and a half long; when creeping about, he stretches more than two inches. The velvet coat is quite gone. In its place he wears a garment softer and smoother than the finest satin, or perhaps more like the delicate kid of which gloves are made, save that the worm's skin is far more delicate. The color is apple-green, paler on the sides, and whitish beneath; the bands are black, dotted with yellow spots, so placed as to form regular lines along his body.

In structure, our caterpillar is an example of all others. His body is made of twelve rings of tolerably firm substance, connected by softer bands, and covered with skin. Thus he has the most perfect freedom of motion. He can stretch or contract himself, can turn or twist in any direction, can roll into a ring, or straighten out stiff, like a twig of the plant on which he feeds, or conform to any unevenness of surface over which he may creep. His head is covered with a flattened, shelly dish, provided on each side with six minute shining grains, which naturalists say are eyes. They do not say that caterpillars can see; Dr. Morris thinks "it is very doubtful whether they have the faculty of vision." One who watches a worm feeding, moving about, reaching out this way and that, quite ignorant of any danger that threatens, passing at the shortest distance the very thing which it seems to seek, never recognizing any thing except what it touches, and shrinking only when it is touched, can scarcely fail to conclude that, however many eyes the worm may have, it is, in fact, quite blind.

The mouth is armed with a pair of strong jaws, which open and shut, not vertically, like those of a dog, or a man, but sidewise. In the middle of the broad under lip is a small elastic tube, with a minute opening, whence comes the silk which it will some day find useful. In tropical countries the head is often queerly ornamented with spikes, prickles, horns, and other things; those which we may see rarely have any thing of the kind.

Each of the first three rings of the body has a pair of jointed, tapering legs, covered with scaly or horny mail, and ending with hooks. These are the true legs. The worm has, besides, four to ten — usually eight — false, or pro-legs. These are thick, fleshy, without joints, but can stretch or contract like the body, are furnished at the end with a fringe of small hooks, and can take very different forms, as the animal wishes to cling by them to various surfaces. Caterpillars which have the full number of legs, that is, sixteen, have still four rings unprovided, the fourth and fifth, and the tenth and eleventh. The twelfth, or anal ring, has always a pair; the ninth has usually a pair; the other pro-legs vary with the species.

The motions of a large caterpillar which has the full complement of legs are deliberate and regular. First he stretches out the elastic body, and puts down the six horny

legs together; then the pair of anal legs take
themselves up, and replace themselves close
behind the pair of the ninth ring, shutting
down upon the twig or leaf, as if made of India
rubber; then the other pairs of pro-legs lift and
move forward, the hindermost rising and falling first, and the others following in their
order; mean while, motion seems to begin at
the tail, and flow gradually and equably through
the entire body, ending by pushing the head
on for another stretch. The motion of such as
have but one or two pairs of pro-legs is similar
in fact, though different in appearance. The
hind legs are drawn forward, and set down just
behind the true legs, the body being thrown up
into a loop; this loop straightened out, carries
on the fore legs again. These caterpillars are
called loopers, geometers, or measurers, since
they seem to measure off the distance of their
journeys. Gail Hamilton's gardener says they
do so: measuring with his thumb and finger on
his coat sleeve.

The looper caterpillars can not shorten or

lengthen their bodies like others, but only
bend them. Some are round and stiff, of the
same color as the bark on which they live.
They grasp the stem or twig with their four
pro-legs, while the body stands out stiff and
motionless for hours together, and the observer mistakes them for twigs, or leaf-stems.

Each kind of caterpillar feeds by choice only
on certain kinds of food, and most will refuse
any other variety. They usually prefer leaves;
after that, flowers; a few eat the pith of the
stalk, and occasional species, the pulp of
fruits. Most feed by night, and remain quiet
by day, as if torpid; some are so voracious as
to eat constantly. A silkworm devours its own
weight of mulberry leaves, daily. Reaumur
gave to a kind which eats cabbage, bits of cabbage leaf which weighed twice as much as their
bodies. The pieces were consumed in less than
twenty-four hours, while the worms increased
their weight one tenth. What if a man weigh-
ing 150 pounds, should eat 300 pounds of food
in a day, and gain 15 pounds of flesh!

When a caterpillar wishes food, it creeps out to the edge of a leaf, and twists its body into such a position that this edge passes between its legs, which hook on upon each side. It bites a mouthful from the edge, then another, and another, moving its head in the arc of a circle, and cutting in three or four bites, as a reaper would cut handfuls of grain with his sickle; the head moves back to the edge of the leaf, and begins another sweep; the fore legs move slowly on from time to time, until the caterpillar has stretched its body to its full length. Then the body draws itself back again, the pro-legs keeping their places, and the head cuts in again for a new swath. The pulp of the leaf is eaten down to the ribs, and often ribs and all disappear between the voracious jaws.

But we must return to our caterpillar of the carrot-leaves. When he has finished eating, he becomes uneasy. He no longer rests quietly on his leaf, or he moves only to find fresh pasturage; he begins to wander about, and if we do not shut him up, we shall lose him altogether. Presently we find him quiet again in some secluded corner at the top of the case; if he could, he would have found a retreat in a knot-hole, a crevice between boards, or an obscure nook under the fence rail. He now presses the elastic tube of his under lip to the wood; the silk material adheres to it; he draws his head away, and stretches a fibre of silk to another point, where he fastens it by pressing the fresh material against the surface. He crosses and recrosses the threads until he has covered a little space with a hillock of silk, to which he fastens himself firmly by the hooks of his hinder feet. Now clinging by his pro-legs, he bends his head back to about the fifth ring, and fastens a thread to the wood beside him. This thread he carries over his back, and fastens on the opposite side; he lays beside it a second, and a third, and in a little time has spun a stout band or loop of silk, in which he may rest securely.

Some caterpillars, like the dark-colored worms, covered with spines, which infest the hop vines, do not spin the band for the back, but content themselves with the little mass of silk into which the hinder hooks are fastened. These simply hang themselves up, and let their bodies fall into a vertical position. The next business is to throw off, for the last time, its skin. To do this, it constantly bends and straightens its body, until the dried skin splits along the back, and part of the body beneath appears. Next, it draws the fore part of the body out of its covering. Then it lengthens and shortens itself by turns, each time splitting the skin still further, and pushing it, like a stocking, nearer to its tail, where it is soon a mere crumpled packet. Now comes the most difficult part of the whole. Out from its caterpillar skin the creature has come in a smooth, horny armor, laid in rings about its body, while its head, back, and breast, are swathed, like a mummy, in folds which firmly confine every limb. It can only wriggle, jerking itself from side to side. Its tail is yet in the folds of the caterpillar skin, which is hooked to the silk above. It must draw itself out of this remnant, throw away the cast off garment, and hook itself by its tail to the same place. We see now the utility of the silken band of our worm of the carrot leaves, but the hop worm has no such assistance. It has neither arms or legs — how can it do so much without losing its own hold, and falling to the ground?

The supple, contracting rings which cover its own body are the limbs which it uses. It seizes a portion of the skin between two of these rings, and so holding on, it curves the tail until it draws it entirely out of the sheath which covered it. But its body is shorter than before this change, and it must climb to reach the tuft of silk to which it should hang. It stretches its body as far as it can, and seizes the skin higher up, between two other rings, at the same time letting go below; this process it repeats with different rings in succession, until finally it reaches the tuft of silk, and fastens to it the hooks in its tail.

It now gives itself a jerk, which sets it to spinning rapidly; it rubs against the skin, and loosens its hold on the silk. If one whirl is not enough, it whirls again, in the opposite direction, and this time will almost surely succeed. Reaumur saw one which, after several efforts to dislodge the old skin, was forced to leave it where it was so firmly fastened.

In about thirty hours after our caterpillar has made himself fast, he has effected this change, and now hangs by his tail, or in his hammock, a pupa, or chrysalis. Here he will remain in unconscious security, during all the quiet days of autumn, and through the bitter blasts and piercing frosts of winter, until the warm breezes of another June awaken his dormant powers to a new life.

Other caterpillars make for themselves cases, or cocoons, spinning them of silk, and often working in other materials. They are for the most part oval, or egg-shaped, sometimes boat-

shaped, and are usually white, yellow, or brown
in color. In some, the threads cling very
slightly; in others, they are closely gummed
together; some are single, others double; some
so closely woven as to quite hide the pupa within, others so thin that it may easily be seen;
some bind together leaves, within which they
hide; some work into the shell bits of earth;
while some weave into the fabric the hairs
with which their own bodies had been covered.

One variety pulls out its hairs with its teeth,
lays them against the web already spun, and
then fastens them by spinning more silk over,
or, rather, under them — for the outside of the
cocoon is spun first, and thickened from within.
Another does not pull out its hairs; it cuts
them off. Another works its hairs through the
meshes of the silken net, and then wriggles
about until it rubs them off. Another pulls
them out in the first place, then sets them up
like the stakes of a palisade, and spins a light

web within, curving them inward so as to form a sort of cradle.

Many caterpillars go into the ground to become chrysalides; there they make round or oblong cocoons. These are always smooth and shining within, and are often fitted with a lining of silk. Reaumur took a cocoon out of the ground, broke it open, and placed it in a glass case containing nothing but sand. In four hours the injury was repaired.

The caterpillar began by coming almost entirely out. It moved its head forwards until it could seize a bit of earth, which it drew into the cocoon; then it came out for another, and so wrought for an hour, gathering material. Then it began to rebuild the broken place. First it spun a band of loose web over a part of the opening; then it placed a few of its grains of earth in the meshes which it had made; it spun more silk, and put more grains in place, binding them together with silken cords. Presently the whole was closed except one small opening, which it filled with crossed threads, and then finally stopped by pushing among the threads the bit of sand which it had saved for the purpose, and which made all tight.

A caterpillar found on the oak trees cuts off thin strips of bark, which it builds into two compact blades; these it so arranges as to form a hollow cone, or boat-shaped shell, in which it becomes a pupa. It is at once architect, cabinet-maker, and weaver.

In due time — sometimes in a few days, sometimes not until another summer, and in one instance, after as many as seven years — the time comes for the last, and most glorious transformation. The poetical Greeks found in this change a type of the liberation of the soul from its mortal tenement, and its entrance into a higher and happier life; hence they called the Butterfly, Psyche, the soul. This idea is most natural. The worm seems to spin its own shroud, to make its own coffin, often to enter its own grave. Yet within this shroud, this coffin, this grave, it lives, a dormaut, waiting life, until the day comes for its resurrection. Then it bursts its cerements, and emerges in a new and beautiful garb, into a brighter existence. But the new life, unlike that of the soul, is brief and mortal; a few short days complete its round, and it perishes forever.

The pupa-case is dry, brittle, and easily broken. The least movement of the fly within opens the dry skin over the middle of the upper part of the thorax; the split extends over the forehead; the pieces separate, and the insect finds an opening through which it may escape. But the escape requires time, for the head, the antennæ, the wings, the legs, sometimes even the tongue, are each in a separate case, and must be liberated one by one. All the parts are soft and moist. The wings, especially, are a pair of crumpled packages, fastened to either side of the thorax. Gradually they unfold, they expand; the insect clings to a twig, and suffers them to hang in such a position that they may expand the more freely; in

time they become dry and firm. If the pupa
is in a cocoon, there is yet more to be done,
for it is still within the silken envelope. In
some, as in the *Cecropia* moth, the end of the
cocoon opposite the head is only partially closed,
and the moth more easily creeps out. Others
cut their way through the silk, for which, Reaumur says, they use their compound eyes as files.
Others exude a liquid which softens the silk,
and assists their escape.

The perfect insect has four wings, covered
with minute scales of varied forms; these,
under the microscope, glow with the most beautiful metallic tints. "Suppose a painter could
present on his canvas, in all their splendor,
gold, silver, the ruby, the sapphire, the emer-
ald, all the precious stones of the East, he
would use no color, or shade of color, which
might not be found on some scales of some
Lepidoptera, where nature has concealed them
from our gaze."

The thorax, or chest, is strongly made, in
order that it may give support to the wings,

and to the six legs. Many have the legs of equal length, and use all in walking; in others, the two fore legs are very short, and are kept folded back against the chest. The body is long, oval, composed of five rings, joined by membrane. The head is rounded, flattened in front, and furnished with hairs. The globular eyes consist of a great number of facets, on which, in different species, glitter all the hues of the rainbow. In the compound eye of the *Papilio,* more than 17,000 facets have been counted. The antennæ are placed near the upper border of each eye. Reaumur has figured six different shapes, and upon them the classification into families partly depends. What is their use? Certainly not for sight, taste, or smell. They are of little use as feelers, and there seems to be nothing else for them to do, which we can understand, except to serve as ears.

The jaws of the caterpillar have disappeared. Instead, the Butterfly has a long, flexible trunk, which it coils up into a small spiral, and carries in a cleft just between the eyes. In some species of Hawk-moths, the tongue is longer than the whole body. It consists of three hollow tubes, a small one placed between two that are larger. Through it the insect draws honey and the juice of flowers. But how can it eat even the most solid sugar? On examination it appears that it sends down through one or two of the tubes of its trunk a fluid which dissolves the honey, or sugar, which is then carried back through the other tube.

After the Butterfly has found its mate, it lays its eggs, some hundreds or thousands in number, upon the plant which is the proper food for its young. They vary much in shape and color. Usually they adhere by a gummy substance; sometimes they are covered with the down from the abdomen of the mother, to protect them from cold, or injury. Some species place them in clusters; others scatter them, leaving only a few upon any single plant. In a few warm days they are hatched, producing minute caterpillars, and the round of nature's course is completed.

Of the *Lepidoptera* some fly by day, others in the twilight, others still in the darkness of night.   Hence authors have classed them as diurnal, crepuscular, and nocturnal.   But this division is not found to be entirely useful, since some that fly by night fly also during the brightest and hottest sunshine, while even the night flyers do not fly all night.   There are three principal sections.

First, there are the Butterflies.   These fly by day, have club shaped antennæ, and when at rest, the fore wings in some, and all the wings in most, stand perpendicularly, turned back to back.

Second, the Hawk-moths.   These fly, some by day, but most in the morning and evening twilight; they have the antennæ thickened in the middle, the wings narrow in proportion to their length, and confined together by a bunch of stiff bristles on the shoulder of the hind wing, which is held by a hook beneath the fore wing; the wings, when at rest, are more or less inclined like a roof, the fore wings covering the under ones.

Third, the Moths. These fly mostly by night. The antennæ taper from the base to the end, and are naked, like a bristle, or feathered on each side; the wings are held together by hooks and bristles, the first pair, when at rest, covering the under pair, and more or less sloped.

Our space will not allow us to describe any of the many varieties of Butterflies and Moths which fly among us. The worm whose changes we traced from the carrot-tops, produces a large, fine Butterfly, called *Papilio Asterias*, which expands from three and a half to four inches. Its color is black; it has a broad band of sulphur-yellow spots across the wings, and a row of fainter yellow spots along the edge. The hind wings are tailed, and have seven blue spots between the two rows of yellow, and an eye-spot of orange, with a black centre.

How to Catch and Preserve Butterflies.

Any active, careful lad can secure a beautiful collection of Butterflies and Moths, in a single season. For this he needs: a net; an ounce of chloroform, or sulphuric ether; pins; a setting-box; suitable boxes for keeping and displaying specimens.

Mosquito-netting is good enough for the net; make a bag about two feet long, and wide enough to be sewed to a light wooden hoop, twelve or fourteen inches in diameter, and fastened firmly to a handle about three feet long. Or, take three or four springs from a discarded hoop-skirt; leave the cotton covering on; slip them through a hem made at the mouth of the net; have them project three or four inches beyond the hem at each side; break off the extra length, and then bend the projecting portions to a right angle; lay these pieces flatly against the handle and bind fast with smooth twine. The net thus made is very light, flexible, and convenient.

For a setting-box, any roomy box, of wood or pasteboard, two or three inches deep, will do. The bottom may be covered with thin sheet cork, pasted or glued down ; or, instead, strips of corn-stalks serve to hold the points of the pins very well. Some strips should have a groove between them one-fourth of an inch deep by three-eighths of an inch wide, to receive the bodies of the larger specimens.

Where nothing better can be had, the bottom of the box may be arranged thus : get strips of inch board, about an inch wide, and as long as the box ; if the edges are sawed smoothly, do not plane them, but smooth the upper surface, and plane off each upper corner ; place the sawed edges of two strips together and nail them ; then nail on a third, and so on, until a board is built wide enough to cover the box. The corners which were planed off now leave triangular grooves, while the sawed edges, though quite close, still allow the pins to pass between them. It is better to nail the strips together than to fasten them with cleats, because the joints hold the pins better. By a little care the grooves may be made of different depths, to receive specimens of different sizes.

Common brass pins may be used; needles of various sizes are better; best of all, the German pins made for the purpose, and sold by dealers in philosophical instruments.

The permanent cases are best of wood, tightly made, and glazed on one or both sides. When only one side is glazed, the bottoms may be fitted like that of the setting box, and should be lined with white paper. Bits of camphor should be fastened in them to drive away insects, or some fine day only a few wings, legs, and the dust of bodies will remain of the most valued specimens.

But little can be gained by striking at Butterflies on the wing. Find one which is resting on a flower, or on the ground; approach quietly, bring the net up carefully until quite sure of him, then turn it skillfully, and he is caught; hold up the bag, while the hoop is flat on the ground, the insect usually rises into it, and the folds falling over prevent the spoiling of the wings.

Now touch the head with a drop of ether, to stupefy him, take him out gently, put a pin through the thorax between the roots of the wings, and place him in the setting-box. As Izaak Walton says of using a frog for bait, "Use him as though you loved him." Arrange the feet as naturally as possible; then with a needle push the fore wings forward until their hinder edges lie nearly in a straight line — beginners do not usually bring them forward enough. Then lay over the wings on each side a strip of paper, or of card, and fasten it down at each end with a pin, which must not pass through the wings. Take care that the two sides are placed alike. Some specimens of each kind should be set upon their backs, to display the under surfaces. Leave them in the setting box until *thoroughly dry*, allowing two or three weeks for the larger kinds; otherwise the wings will get awry, or droop, and the whole have an awkward appearance.

Ether, and chloroform, often fail to kill; some of the larger moths take large and repeated doses, and still live. For such, a little cyanide of potassium may be had, dissolved in water. A drop taken on a needle and pricked into the thorax under the wings, is merciful to the poor captive. Great care must be taken with this substance, for it is very poisonous when taken into the mouth.

Hawk-moths, and many others, fly very swiftly, and require great dexterity in their captor; take them when busy with a flower. Many moths may be attracted through an open window with a light. During the day they may often be found resting, head down, on fences, bark of trees, and elsewhere. Cover your specimen with a glass, slip a paper under, and take him away; a few drops of ether on the paper fills the glass with vapor, which suffocates the insect. Some of each kind of moth should be set up, with wings in the natural position, as when at rest.

Set up a number of specimens of each kind, in order to secure a choice; four are always wanted to show the upper and under surfaces of both male and female. Reject at once all that are broken-winged, or otherwise injured, unless the species is rare, and then as soon as a better one is found. The collector secures his finest specimens by saving the cocoons, and taking the flies as soon as hatched, before they have had time to injure themselves. The cocoons should be kept through the winter in a cool place, in a roomy box; when the time comes for hatching, twigs must be provided, on which the butterflies may rest while the wings are expanding, else they may be hopelessly crippled.

Early in summer, get a candle box, and raise the lid about twelve inches, on strips of board nailed into the four corners; cover three sides of the open case with wire gauze, and fit a door to the fourth side. Fill the box with fresh garden mould, and set it away in a shady place. If a new caterpillar is found, put him in the case with *plenty* of *fresh* food. The inhabitants will not quarrel, and will usually thrive. When grown, some will descend into the earth; some will spin cocoons, and some will hang themselves up in the corners. Keep through winter in a cool place, away from mice, and watch the coming out of the insects in the spring.

A little patience and contrivance will do all we have described, and more, while much pleasure and instruction will be gained. Even this, profitable though it may be, should not be allowed to interfere with the performance of regular duty, whether work or study.

NOTE.—The general reader who desires further information concerning the species and habits of insects, will find "Harris' Insects Injurious to Vegetation," and the "Guide to the Study of Entomology," by A. S. Packard, Jr., now issuing in numbers in Salem, Mass., best suited to his purpose.

www.ingramcontent.com/pod-product-compliance
Lightning Source LLC
Chambersburg PA
CBHW021941220326
41599CB00013BA/1470